A Monograph on

SHEATH ROT DISEASE OF RICE

SHEATH ROT DISEASE OF RICE

K. MANIBHUSHAN RAO
Centre for Advanced Studies in Botany
University of Madras, Guindy Campus
Madras 600 025, India

1996

DAYA PUBLISHING HOUSE
DELHI-110035

Front Cover : Typical sheath rot symptoms on rice (L-R): oblong, dark-brownish lesions on the leaf sheath: young affected panicle partially emerged and the glume with dark-brown to chaffy grains respectively.

ISBN 81-7035-163-4

Published by : **Daya Publishing House**
1123/74, Deva Ram Park
Tri Nagar, Delhi - 110035.
Phone : 7103999
Fax (011) 7198902

Laser Typeset by : ShiKi Associates, New-Delhi - 110058

Printed at : DP's Impressive Impressions, New Delhi - 110059

PRINTED IN INDIA

. . . . .to Annapurna, Sreenivas, Sudhir, Padmaja, Ramakrishna and Harsha

Acknowledgements

I have great pleasure in thanking the authors, publishers and Institutes to reproduce plates, figures, tables and copy right © materials used in the text.

Dr. P. Lakshmanan for fig. 1; Dr. D.L. Hawksworth for figs. 3ab, 4ab, 6, 7&8; Dr. V. Agnihothrudu for fig. 5abc; copyright © 1995, Eugen Ulmer Gmb H & Co, Stuttgart for fig. 10 and tables 8 and 10; Dr. M.S. Kang for plate 2; Dr. A.K. Deka for table 11; Dr. S. S. Gnanamanickam for plate 6 and table 15 (copyright ©1991) reprinted by permission from Kluwer Academic Publishers, The Netherlands; Dr. S.S. Gnanamanickam for plates 7 and 8; Dr. R.P. Purkayastha for figs. 12,18 & 19; and Dr. K. Satyanarayana for fig. 15; Dr. S.I. Akanda for fig. 16. Bibliographer, International Rice Research Institute(IRRI), Manila, The Philippines for fig. 20. Finally my erstwhile colleagues and research students for reproducing several of joint publications which are listed appropriately in the text.

Foreword

Sheath rot (ShR) is considered to be one of the serious diseases of rice in the South and the South-East Asia. The disease became widespread with the introduction of the high yielding N-responsive rice cultivars throughout this region, including the Indian sub-continent.

Professor K. Manibhushan Rao with his rich experience in rice research of over three decades has made commendable efforts in bringing out the present monograph which summarizes the state of the current knowledge in this field. Prof. Rao has dealt with all aspects of the ShR disease including host-parasite interaction, resistance, biology and ultrastructure of the causal organism and its association with other biotic stresses such as the rice-tungro virus, mites, stem-borers and mealybugs. The chapters dealing with disease control and future perspectives cover a wide range of topics from the mode of action of fungicides to the use of antibiotics and biocontrol agents suggesting several new research possibilities for the future.

The monograph fills an important gap in our knowledge of this disease. I am sure that this critical treatise will not only serve as a useful reference but also stimulate further research on disease management in rice.

R.S. Paroda
Secretary & Director-General
Indian Council of Agricultural Research
Ministry of Agriculture,
Krishi Bhavan, New Delhi-110001

New Delhi
May 10, 1996

Preface

Rice (*Oryza sativa* L.) is an economically important crop cultivated in more than 147 million hectares world wide. The current global production is 520 million tonnes. It is estimated that in the next 3 years, rice production must increase by 65 per cent to keep up the present population growth. However, the rice yield is most often adversely affected by the diseases caused by plant pathogens, insects and pests. Blast(Bl), sheath blight (ShB) and bacterial blight are considered as the major rice diseases while sheath rot (ShR), as a minor one. The introduction of high yielding, semi-dwarf, nitrogen responsive cultivars (cvs), compounded with the conducive microclimate resulting from close planting method, increases the ShR disease incidence in recent years and poses a major threat to rice cultivation in the tropics.

I have been working on rice Bl and ShB diseases over three decades, and had the good fortune in bringing out the respective monographs in 1994 and 1995. Besides working on these two principal rice diseases, I have initiated studies on ShR in early 1980's from different angles with the cooperation of my research students and the significant developments made were consolidated and a comprehensive review was published in 1986. Subsequently, the literature on ShR research was updated and the advances made were discussed in the present monographic study.

Further, the current perspectives, the challenges for the future as well as specific problems that have surfaced were outlined. The nomenclature and taxonomic position of the pathogen *Acrocylindrium oryzae* (=*Saracladium oryzae* syn. *S. attenuatum*) was in a moot point and based on the phytopathological and morphological relevance etc. this was resolved and grouped separately. There is little information available on ShR during 1965-75 and the pertinent literature is available from mid 1980's only. In recent years much progress has been made on the ShR disease and I have included the latest available information, a factor, I consider is very important in writing this monograph and also for world wide usage by the pathologists, students and extension workers as well as rice researchers.

The unpublished material from my student's, M.Phil. and Ph.D. dissertations especially from Mrs. V. Parvathy, Mrs. A. Malathy and Dr. Y. Joe respectively were appropriately quoted. I am grateful to Prof. P. Vidhyasekaran, TNAU, Coimbatore for reading the entire MS and for his valuable comments and suggestions. However, I own complete responsibility for any misinterpretations, omissions, deficiencies that may remain. I shall always welcome any further comments and suggestions for the improvement of the book from rice researchers.

The encouragement and good wishes from my wellwishers prompted me to take up this critical study followed by a good response from rice researchers in my earlier publications on "Rice blast disease" and "Rice sheath blight disease" in 1994 and 1995 respectively. My visit to Japan recently with INSA-JSPS collaborative scientific exchange programme to various institutes/universities helped me in updating the literature available on this subject.

I express my profound gratitude to the University Grants Commission (UGC) and Indian Council of Agricultural Research (ICAR), New Delhi, India for extending continued financial support to my research projects on rice diseases over these years. I appreciate the interest of M/S Daya Publishing House, Delhi, for taking up this book at such a short notice. I thank the University of Madras for its support and Prof. A. Mahadevan, Director, CAS in Botany for facilities. I will be failing in my duty if I have not mentioned the cooperation and help I received from my wife Mrs. Annapurna and all my family members.

K. Manibhushan Rao

Contents

Standard Abbreviations

are(s)	:	100 sq. metres
°C	:	degree Celsius
cm	:	centimetre(s)
cv.(s)	:	cultivar(s)
DAS	:	days after sowing
DAT	:	days after transplanting
dia.	:	diameter
g	:	gram(s)
g^{-1}	:	per gram
h	:	hour(s)
ha	:	hectare (s)
kg	:	kilogram(s)
l	:	litre(s)
m	:	metre(s)
ml^{-1}	:	per millilitre
µ	:	micron($x10^{-6}$)
m^2	:	square metre(s)
mg	:	milligram(s)
min	:	minute(s)
ml	:	millilitre(s)
mw	:	molecular weight
N	:	nitrogen
p.(pp.)	:	page(s)
pH	:	hydrogen ion concentration
ppm	:	parts per million
RH	:	relative humidity
sec.	:	second(s)
sp.(p)	:	species
t	:	metric tonne(s)
wk	:	week
wt.	:	weight

±	:	with or without
%	:	percent(age)
/	:	per
α	:	alpha
ß	:	beta

History and Distribution

Sheath rot (ShR) of rice caused by *Sarocladium oryzae* (Sawada) Gams and Hawks* was first reported from Formosa, Taiwan, by Sawada (1922). The occurrence of the ShR has also been recorded from Japan (Kawamura, 1940; Tasugi and Ikeda, 1956), Vietnam (Anonymous, 1962), Thailand (Hashioka, 1963; Ou, 1973 and Surin and Disthaporn, 1977), United States (Attabhanyo and Rush, 1973, Shahjahan et al. 1977 and Harahep et al. 1973), Peru (Jimenez and Carbos, 1977), Brazil (Mathur, 1981) and Bangladesh (Rahman et al. 1982), Nigeria (Imolehin, 1989) and Punjab in Pakistan (Gill et al, 1993). In India, the occurrence of ShR was reported for the first time by Agnihothrudu in 1973. Further, Amin et al. (1974) and Reddy et al. (1974) reported the ShR incidence from Andhra Pradesh. Though ShR was considered as a minor disease and its wide-spread occurrence was recorded from several rice growing states of India, viz., Andhra Pradesh (Reddy et al. 1974; Koteswararao, 1977 and Muralidharan and Venkatarao, 1980), West Bengal and Bihar (Amin et al. 1974; Pai, 1975 and Ghufran et al. 1980), Tamil Nadu (Prabhakaran et al. 1974, Ahmed et al. 1975 and Lakshmanan, 1991), Orissa (Naik and Roy, 1975, Nair, 1976), Kerala (Nair and Rajan, 1978), Punjab (Raina and Singh, 1980; Kang and Rattan, 1983), Uttar Pradesh (Raju and Singh, 1981), Rajasthan (Tripathi, 1982) and Madhya Pradesh (Upadhayay and Diwakar, 1984). Chakravarty and Biswas (1978) and Shahjahan et al. (1977) reported 14.5% and 11.4% of ShR incidence in India and the United States respectively.

With the introducion of high yielding cultivars (cvs) severity of the ShR disease incidence increased and therefore, it poses a potential threat to rice cultivation in tropics. Narain (1991) reported about a few minor diseases including ShR posing threat to rice cultivation in India. Oflate, it has been recognized as one of the major rice diseases next to blast (Bl) and sheath blight (ShB) (Manibhushanrao, 1994; 1995). In Tamil Nadu, all high yielding rice cvs are found to be susceptible for ShR which is severe during late *kharif* (August-February) season (Lewin and Vidhyasekaran, 1987b). Further, Buddenhagen (unpublished) also considered it a major problem in upland rice cultivation in West-Africa on rice cvs introduced from Asia.

Earlier, this malady of rice has been briefly mentioned by Padwick (1950) and later by Ou(1985). Further, a comprehensive account of the rice ShR disease and its control has been reviewed by Ghosal and Purkayastha (1981), Purkayastha and Ghosal (1982) and Manibhushanrao et al. (1986). Growth (1992) gave an extensive coverage of ShR in the compendium of rice diseases.

---

* In the earlier literature *Acrocylindrium oryzae* Sawada was used as the causal organism of ShR disease by several rice research workers and the author in a number of publications. However, in the recent literature, both the species, viz., *Sarocladium oryzae* and *S. attenuatum* were used and retained when such work is referred in the text.

Symptomatology

Sheath rot of rice which leads to the production of chaffy grains in the panicle is popularly and appropriately known as "empty head" and is familiar as "rice abortion' among Thai farmers. Initially, the lesions appear as oval or irregular spots with greyish centres and brown regular or irregular margins surrounded by a pale brown halo on the flag leaf sheath.

The pathogen primarily attacks the flag leaf sheaths through stomata or injuries caused by insects and ramifies intercellulary in the vascular bundles and mesophyll tissue of susceptible cvs (Shajahan et al. 1977). In most cases the affected plants become sterile (Hsieh et al. 1977). The ShR symptoms also appear as spotting of the husks as well as sterile spikelets (Buddenhagen, 1985). The pathogen causes serious discoloration on empty grains (Arunyanart et al. 1981) and the grain size will be affected in the ShR injected plants (Chen and Chein, 1964).

The disease symptom appears on the upper most flag leaf sheath, enclosing the emerging young panicles as oblong to irregular brownish lesions (0.5 to 1.5 cm length and 0.3 to 0.5 cm width) with an undulate dark-brown margin with greyish centres which later enlarge and coalese (Tasugi and Ikeda, 1956). The fungus may be seen forming superficial mycelium with spore mass in the centre of the lesions and occasionally inside the rotted sheaths. The severely affected panicles do not emerge from the leafsheath and this effect is known as 'choking'. The occurrence of choking and ShR symptoms with discoloured grains is common on dwarf cvs but varied with cvs, locations and seasons (Amin et al. 1974).

The symptoms caused by *S. oryzae* and *S. attenuatum* are indistinguishable from each other (Brady, 1980a,b). The field grown rice plants (susceptible cv. CO 43) showing typical ShR symptoms (Plate 1 a,b) by *S. oryzae* and *S. attenuatum* on flag leaf sheaths and glumes show dark-brown to chaffy grains (Joe and Manibhushanrao, 1996a).

a b

Plate 1ab. Field grown rice plants (cv. CO 43) showing typical ShR symptoms (arrows) on flag leaf sheaths and glumes a) *Sarocladium oryzae*, b) *S. attenuatum*. Note: The dark brown (db) and chaffy (ch) grains (Joe and Manibhushanrao, 1996a).

Disease Development

The pathogen enters through stomata or any external injury, and colonize the host tissuse intercellularly (Shahjahan et al. 1977). The fungus attacks the tillers at random at all stages but most often causing maximum destruction at booting stage. The disease intensity ranges from partial to complete coverage of the panicle within the flag leaf sheath. Other associated characteristics of the disease are grain discolouration and sterility (Estrada and Bonman, 1984).

Hashioka and Makano (1956) reported that a positive correlation was established between N-fertilization of rice plants and the disease development of ShR. However, reports on the influence of N-fertilization on host susceptibility are conflicting.

The infection process and spread of the disease is mainly on the susceptibility of the host and virulence of the pathogen. Information on the disease development, spread and host colonization of the pathogen, are quite essential. Many of these related terms are mostly used in the assessment of plant diseases. Detailed information on such terminology used in relation to rice blast (Manibhushanrao, 1994) and rice sheath blight (Manibhushanrao, 1995) diseases has been discussed.

The pathogen is seed-borne and survives in soil on the leftover stubbles, which are the sources of primary inoculum (Reddy et al. 1974). Several weed hosts (Balakrishnan and Nair, 1981; Rahman et al., 1982) appear to play an important role in the perpetuation of the fungus resulting in the spread of disease. Rice tungro virus (RTV) affected rice cvs Srinivas, NLR 13969 and Rasi showed severe ShR incidence of 30, 36 and 17% respectively, while RTV free cvs showed only 5-8% ShR incidence (Naidu and Reddy, 1989). Earlier, Venkataraman et al. (1987) reported that RTV infected plants showed mild to severe ShR symptoms of ShR disease whereas the difference was discernible without RTV infection.

ShR infection primarily occurs at the booting/flowering stage of rice crop. Although the upper leaf sheaths are infected, the disease is most conspicuous on the flag leaf sheath enclosing the young panicle. Further, *S. oryzae* could infect rice plants at all growth stages and the disease becomes severe at late maturing stage.

Studies carried out so far on the effect of environmental factors on ShR development are inadequate to give a clear picture. Severe ShR disease incidence has been reported by Koteswararao (1977) and Shahjahan et al. (1977) in *rabi* (dry season i.e., November-March) as well as *Kharif* (monsoon season i.e., June-October). Further, Upadhyay and Diwakar (1984) and Gopinath (1986) have reported a temperature and humidity range of 20-30°C and 65-80% respectively for the optimal disease development. Nevertheless, Singh and Raju (1981) could not find any correlation between temperature, rainfall, sunshine and the disease spread. Sheath rot which co-exists with the rice crop almost throughout the year seems to prefer the prevalence of cloudy weather particularly during the flowering stage of the crop.

Disease Incidence

Disease incidence of rice ShR is associated with insect injury or RTV viral infection (Ou, 1973; Chin, 1974 and Reddy and Ghosh, 1985). Further, Chen and Chien (1964) and Amin et al. (1974) and Ou (1985) found stem-borer injury or viral infection on more than 50% of the ShR infected rice plants. Several researchers (Hsieh et al. 1977; Surin and Disthaporn, 1977; Chen and Huang, 179; Lo and Ho, 1979; Chein, 1980; Chow et al. 1980 and Liang, 1980) confirmed the biological association between *S. oryzae* and the mites (*Steneotarsonemus madecassus* and *S. spinki*) and also the role of these mites in disease spread and severity. The tarsonemid mite, which punctures the soft surface of the epidermis on rice leaves and sheaths, sucks the juice and also transmits *S. oryzae* spores carried on its body. It is an important pest of rice and is always associated with sterile rice plants. Fang (1980) observed a decrease in *S. spinki* population, number of infected and chaffy grains and also an increase in the fertility of rice plants upon treatment with ethyl parathion (47%). Information on the exact cause of rice sterility is not available and yet to be investigated (Purkayastha et al. 1983).

Fungal pathogens, *Fusarium roseum and Hyalostachybotrys* sp. (Shahjahan et al. 1977), *F. equiseti* and *F. moniliforme* (Kang and Rattan, 1983) were co-existent in ShR affected rice tissue. They also attributed the *Fusarium* spp. a role in rice ShR and panicle sterility. Plants were inoculated with conidial suspension or by inserting the mycelial bits into the boot leaf sheaths before panicle emergence gave typical ShR symptoms (Plate 2).

Wei (1957) compiled the list of several fungal species on rice responsible for grain discoloration. While Vidhyasekaran et al. (1983) reported the poor quality of rice grain from ShR infected plants. This is largely due to discoloration and deterioration of the grains. Further, Zeigler Alrarez (1989) reported that *Pseudomonas* spp. cause rice ShR disease as well as grain discoloration. Studies conducted at our centre also showed fast-growing myclelia of fusaria masked the growth of ShR pathogen *S. oryzae* in culture. *Fusarium* spp. were isolated consistently from more than 10 samples of diseased material from South India and induced identical ShR symptoms. Whether the *Fusarium* spp. have any role in ShR disease development or not has to be verified.

a b

Plate 2ab. Typical sheath rot symptoms caused by a) *Fusarium equiseti* and b) *Fusarium moniliforme* on rice leaf sheaths (Kang and Rattan, 1983).

Disease Severity and Yield Loss

Several ShR disease outbreaks have been reported in *rabi* (Amin et al. 1974; Misra and Mathur, 1983) as well as in *Kharif* seasons (Chien, 1980; Srinivasan, 1981). Anima (1975) reported most high yielding cvs are affected with ShR in varying degrees (affected panicles) 4.2-69.1% during *rabi* season for cvs Kanchi and IR 8 respectively while Mohan and Subramanian (1979a) also reported heavy yield loss due to ShR disease. Quantification of rice ShR disease is quite essential for the evaluation of yield losses accurately. Singh et al. (1985) used a simple method for assessment of yield due to ShR.

Sheath rot disease is now considered a potential threat to rice cultivation in both temperate and tropical regions due to severe damage and high yield losses. The extent of yield loss was 3-20% in Taiwan (Chen and Chen, 1964), 52.8% total yield loss in the Philippines (Estrada et al. 1984), 85% to almost total destruction in Thailand (Surin and Disthaporn, 1977) and 11.4% in the United States (Shahjahan et al. 1977) and 14.5% in India (Chakrabarthy and Biswas, 1978). The yield losses reported from different states in India were as high as 50% from Rajasthan and Punjab (Tripathi, 1982; Raina and Singh, 1980; Kang and Rattan, 1983) and 80 to 85% from Tamil Nadu (Prabhakaran et al., 1974 and Alagarsamy and Bhaskaran, 1986b), Andhra Pradesh (Muralidharan and Venkatarao, 1980) and Kerala (Balakrishnan and Nair, 1981). Srinivasan (1980a) and Thrimurty et al. (1980) studied the incidence of ShR and chaffy grain percentage increase on certain popular rice cvs while Estrada and Bonman (1984) reported an estimated 52.8% yield loss. Further, Tikoo (1985) studied ShR pathogen's (*A. oryzae*) effect on yield as a major component. While Surin et al. (1988) observed a positive correlation between disease incidence (DI) and disease severity (DS) ($r^2 = 0.93$) and between DS and yield ($r^2 = 0.63$). The yield loss was calculated as yield of disease free plot - yield of infected plot × 100 yield of disease free plot.

With high DS, the grain loss increased and the significant regression coffecient of percent of yield loss on DS was calculated as: $Y = 4.287X - 0.146$, $R^2 = 0.623$.

The prime concern is due to ShR incidence during the boot leaf stage of the rice plants which will directly affect the panicle initiation and grain filling resulting in heavy yield losses. Quantitative data on flag leaf sheath infection, panicle exsertion, grain filling and discolouration was compared with a visual scale (IRRI, 1981). The results showed a clear relationship between panicle yield and DS. The yields decreased more intensively as DS increases (Table 1). The percentage of filled spikelets affected was most. Yoshida et al. (1976) suggested the grain yield per plot as follows : Grain weight from harvested hills / Total no. of harvested hills × Total no. of hills in the normal plot.

The percent yield was computed using the formulae as $(1\text{-}yd/yh)\times100$; Where, yd = yield of diseased plants; yh = yield of healthy plants and $y = a + bX$; where a = intercept, b = slope of the line and X = DSI (Disease severity index) computed as :

Sum of FLS at each grade No. of FLS scored x maximum grade; where FLS represents flag leaf sheath.

Table 1: Yield components of panicles with various sheath rot (ShR) severity rating (IRRI, 1981)[@]

Severity rating	1000 grain wt.(g)	Spikelets (No./panicle)	Filled spikelets (%)	Relative yield (%)
0	24.9a	164.2a	95.0a	100.0
1	24.3a	178.5a	90.7a	88.3
2	22.3b	171.4a	59.8b	62.2
3	14.8c	144.5b	2.7c	1.3

@ Healthy panicles and diseased panicles of cv. IR 442 selected from the field.

Lakshmanan et al.(1992b) reported the yield loss as well as seedling vigour due to ShR and mealy bug (*Brevennia rechi*) interaction. ShR severity was positively correlated ($r = 0.99$) with the mealy bug population (Fig. 1). Their interaction greatly reduced 1000 grain weight, grain yield (9.77%) and healthy grain percentage while it increased chaffy and discoloured grains. Further, Maiti et al. (1992) assessed the yield loss in upland rice due to ShR using a linear prediction model.

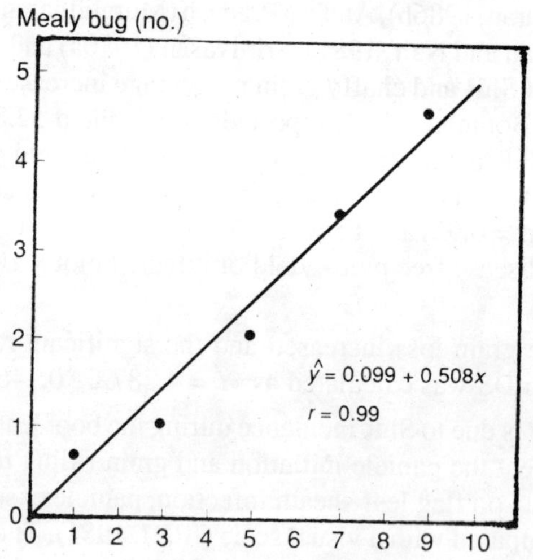

Fig. 1 : Correlation of sheath rot severity and mealy bug population (Lakshamanan et al. 1991)

Joe (1994) studied the relationship between DSI and yield loss (Fig. 2) in two rice cvs resistant and susceptible to ShR, White Ponni and CO 43 respectively. Earlier, Alagarsamy and Bhaskaran (1986b) reported that with the increase age of the plants, the ShR disease severity also increased (Table 2) . Virulence of the isolates also played a major role in causing ShR disease resulting considerable loss in grain yield.

Table 2: Disease severity index (DSI) and yield loss in resistant (White Ponni) and susceptible (CO 43) rice cvs as influenced by N-fertilization and ShR infection (Joe, 1994).

Isolate	Treatment	Resistant		Susceptible	
		DSI	Yield loss(%)	DSI	Yield loss(%)
S. oryzae					
SO 1	UF	31.9	38.6	68.5	45.0
	NF	43.7	54.6	89.9	72.4
SO 5	UF	22.6	36.0	57.4	43.5
	NF	27.5	50.1	63.9	57.3
S. attenuatum					
SA 1	UF	28.3	38.5	62.9	42.6
	NF	32.4	55.7	73.8	52.1
SA 5	UF	18.6	20.9	55.2	24.6
	NF	24.9	46.9	60.3	41.9

UF : Unfertilized; NF : N-fertilized

Linear and Multiple Regression Analyses (MRA)

The relationship between DSI and yield loss in both resistant White Ponni and susceptible CO 43 rice cvs was determined using linear regression (Joe, 1994). The effect of individual factors on the respective cv. was analysed separately by the multiple regression using the following formula:

$$Y = a + b_1 * UF + b_2 * DSI + b_3 * SO1 + b_4 * SO5 + b_5 * SA1 \text{ using SA5 as reference.}$$

The correlation coefficient of DSI vs yield loss of the resistant and susceptible rice cvs when tested individually and revealed variation explained by the DSI for White Ponni cv. was 44% ($R^2 = 44\%$). This predicted the involvement of a few more factors which contribute to the grain loss. On the other hand, in cv. CO 43 the variation caused by DSI towards the yield was 67%, which explained the involvement of a less number of other factors on yield loss (Fig. 3). However, this analysis failed to express the definite factor (N- fertilizer/ isolate/ any other parameters) responsible for yield losses.

Factors affecting yield loss

Joe (1994) reported 82% variation in the yield loss was attributed to the factors like N-fertilizer application and differentially virulent isolates. Presence of N-fertilizer contribute

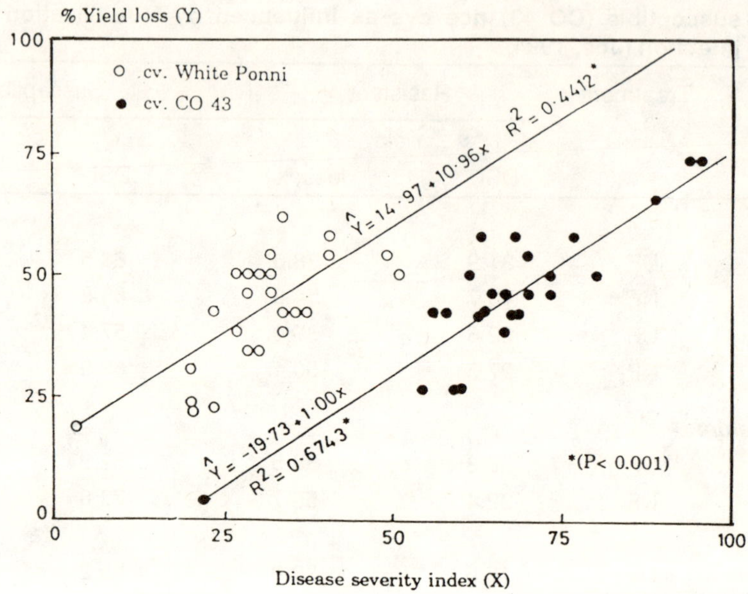

Fig. 2 : Relationship between disease severity index and yield loss (Joe, 1994)

16.9% yield loss which was highly significant (P< = 0.001). Further, isolates SO 1, SO 5 and SA 1 caused 11.35, 8.41 and 9.31% yield loss respectively. Variables like differentially virulant isolates, N-fertilizer and DSI caused 89% variation which conferred a high yield loss on the susceptible cv. CO 43. This isolate SO 1 exhibited 16.12%, SO 5-14.35% and SA 1-8.62% variation compared to the isolates SA 5 towards yield loss. Further, N-fertilizer also had a significant (0.1% level) effect causing 11.6% grain loss (Table 2). In general, this study indicated that an increase in DSI by one unit attributed 0.5% loss in yield in susceptible cv. while in the resistant cv. it was only 0.2%. This further confirmed that the influence of disease was more on the susceptibility than the resistance cvs.

Causal Organism

Acrocylindrium oryzae Sawada: Class: Hyphomycetes; Order: Moniliales; Family: Moniliaceae.

Acrocylindrium oryzae Sawada, Rep. Govt. Res. Inst. Dept. Agric. Formosa, **2**: 1-35, 1922.

Sarocladium oryzae Sawada, W.Gams & D. Hawksw Comb., Kavaka,**3**: 57-61,1975.

Sarocladium attenuatum W.Gams & D. Hawksw sp. nov., Kavaka, **3**: 57-61, 1975.

Fusarium roseum Link., *Hyalostachybotrys* sp. Pl. Dis. Rep. **61**: 307-310, 1977.

Steneotarsonemus madecassus : *S. spinki* Smiley, Pl. Protect. Bull. Taiwan, **19**: 3-36, 1977.

Fusarium equiseti (Corda) Sacc., *Fusarium moniliforme* Sheldon. : Int. Rice Res. Newsl., **8** (3): 7-8, 1983.

Pseudomonas syringae pv. *oryzicola*, Int. Rice Res. Newsl. **14** (1): 26, 1989.

P. fuscovaginae. Int. Rice Res. Newsl., **18** (2): 27-28, 1993.

Curvularia lunata (Walker) Boedijn. Int. Rice Res. Newsl. **16** (5): 20, 1991.

Brevennia rehi Lindinger, Int. Rice Res. Newsl. **17** (2):23, 1992b.

Fusarium moniliforme Sheldon *F.graminearum* Schwabe, *F. avenaceum* (Corda ex Fr.) Sacc. Int. Rice Res. Newsl. **19** (3): 28, 1994.

Nomenclature

The genus *Acrocylindrium* was originally described by Bonorden (1851), and Sawada from Taiwan (1922) described the causal fungus *Acrocylindrium oryzae* Sawada. Much later, Gams and Hawksworth (1975) reviewed the nomenclature and taxonomic position of *A. oryzae*, recognised its phytopathological relevance and transferred to a new genus *Sarolcladium* as *S. oryzae* (Sawada) W.Gams & D. Hawksw. Further, *A. oryzae* isolate on paddy described by Agnihothrudu (1973) was identified as *S. attenuatum* after comparing Sawada's holotype material. This gave the confirmation that *A. oryzae* described by Amin et al. (1974) was identical with Sawada's holotype but the isolate described by Agnihothrudu (1973) was different. Subsequently, the relevance of *S. attenuatum* as alternate/co-pathogen on rice was realized. Recently an attempt was made to study different biochemical constituents of virulent isolates of *Sarocladium* spp. (Joe and Manibhushanrao, 1995).

Taxonomy

The taxonomic position of this genus described by Gams and Hawksworth (1975) is as follows:

Genus hypomycetum *Verticilli* et *Gliocladii* simile, sed phialidibus plerumque scoparum modo secundum conidiophori ramos dense oriundis differens. Phialides anguste cylindricae, vix attenuatae, collari carentes; conidia continua, cyclindrica, hyalina.

Species holotypica *Sarocladium oryzae* (Sawada) W. Gams & D. Hawksw.

Genus of Hyphomycetes resembling *Verticillium* and *Gliocladium.* Colonies spreading with numerous conidiophores arising from aerial or submerged hyphae, irregularly and densely branched with the branches and phialides arising in dense broom-like rows along the supporting cells. Phialides narrowly cylindrical, hardly tapering towards the apices, lacking collarettes; conidia one-celled, cyclindrical, hyaline.

Type species

Sarocladium oryzae (Sawada) W. Gams & D. Hawksw. comb., *Kavaka* 3:57, (1975) (Fig. 3a).

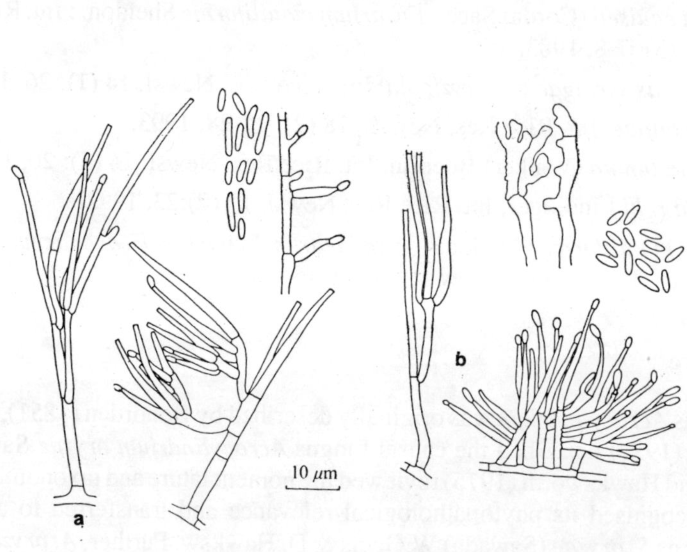

Fig. 3ab : (a) *Sarocladium oryzae* hyphae with condiophores and spores (b) vegetative hyphae with irregularly generated outlines (Gams and Hawksworth, 1975).

Basionym: *Acrocylindrium oryzae* Sawada, *Rep. Govt. Res. Inst. Dep. Agric. Formosa* 2: 135 (1922).

Colonies on the natural host superficial, spreading, pale pinkish or whitish. Colonies on 2% MEA reaching 13-15 mm diam. in 10 d at room temperature, pale saffron, with a thin compacted cottony aerial mycelium, reverse darker orange-ochre. Vegetative hyphae 1-1.5μm diam. wide; in the aerial mycelium some wider hyphae up to 2.5μm diam. with irregularly gnarled outlines (Fig.3b). Sporulation abundant in the central parts of the colonies. Fertile hyphae giving a strong metachromophilic reaction with aniline blue. Conidiophores irregularly branched; sometimes with several branches arising in whorls at one level but more commonly with branches arising laterally in dense rows over a considerable length of the supporting cell; secondary branches often exceeding the primary ones; repeatedly branched conidiophores up to 60 µm or more in length and 2-2.5 µm wide at the base. Conidiogenous cells (phialides) arising from conidiophores or directly from undifferentiated vegetative hyphae, 30-40 µm long and 1.5-2 µm wide at the base when formed at the apex of slender conidiophores, 6-20 µm long and 1-1.5 µm wide when arising in dense broom-like fasicles; tips tapering to 0.6-1 µm in width, lacking any distinct collarette. Conidia formed in slimy masses, cylindrical with rounded ends, sometimes becoming slightly curved, hyaline, thin-walled, smooth, one-celled, 3.5-7 x 0.8 -1.5 µm. Chlamydospores absent.

Sarocladium attenuatum W. Gams & D. Hawksw. sp.nov., *Kavaka* **3**:57, (1975) (Fig. 4a).

Coloniae dilute roseae abunde sporulant. Conidiophora capitula conidiorum discreta proferunt. Conidiophora e stipite brevi continuo vel bieellulari et plerumque singulis vel binis verticillis ramorum appressorum constantia, cellulis terminalibus phialidibus. Phialides solitariae vulgo adsunt. Conidiophora composita 60-90 µm alta 2.5µm diam. ad basam; phialides tenuitunicatae, cyclindricae, 22-45 µm longae, 1.5-2 µm diam. ad basim, 0.7-1 µm diam. ad apicem, collari carent. Conidia fere fusiformia, utrinque truncatula, nonnumquam modice curvata, hyalina, levia, 4.5-8 (-14) x 0.6-1 µm.

Colonies on the natural host superficial, spreading, whitish. Colonies on 2% MEA reaching 12-14 mm diam. in 10 d at room temperature, pale pink with tufts of aerial mycelium; reverse pale salmon particularly at 33°C. Vegetative hyphae 1.5-2.5 µm wide, with wider, gnarled hyphae occurring in the aerial mycelium. Sporulation abundant, mostly with discrete conidiophores bearing slimy heads of conidia. Vegetative basal hyphae and stipes of conidiophores giving a strong metachromophilic reaction with aniline blue. Conidiophores composed of a short stripe of 1-2 cells and usually 1-2 whorls of a few appressed branches, the ultimate branches acting as phialides; solitary phialides also frequently occurring; compound conidiophores 60-90 µm tall and 2.5 µm wide at the base. Conidiogenous cells (phialides) arising from the conidiophores or directly from undifferentiated hyphae, 22-45 µm long and 1.5-2 µm wide at the base; tips tapering to 0.7-1 µm in width, lacking any distinct collarette. Conidia forming in slimy heads, slightly fusiform with truncated ends, some times slightly curved, hyaline, thin- walled, smooth, 4.5- 8(-14) x 0.6- 1 µm (Fig.4b). Chlamydospores absent.

Acrocylindrium oryzae Sawada, Rep. Govt. Res. Inst. Dept. Agric. Formosa, **2**: 135, 1992.

Agnihothrudu (1973) described the ShR pathogen *A.oryzae* on paddy for the first time from India and suggested that the genus might have to be assigned to either *Gliocladium*, or *Clonostachys*. The description made by him is as follows:

On potato dextrose agar mycelium of the fungus effuse, white, sparsely branched, up to 2µm in diameter. Conidiophores arising from the mycelium, up to 3µm in

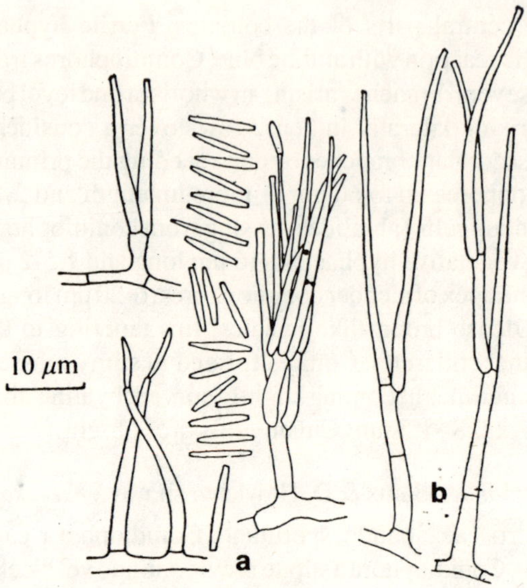

Fig. 4ab : (a) *Sarocladium attenuatum* CBS 339-7-3, (b) CBS 101.61 × 100 (Gams and Hawksworth, 1975).

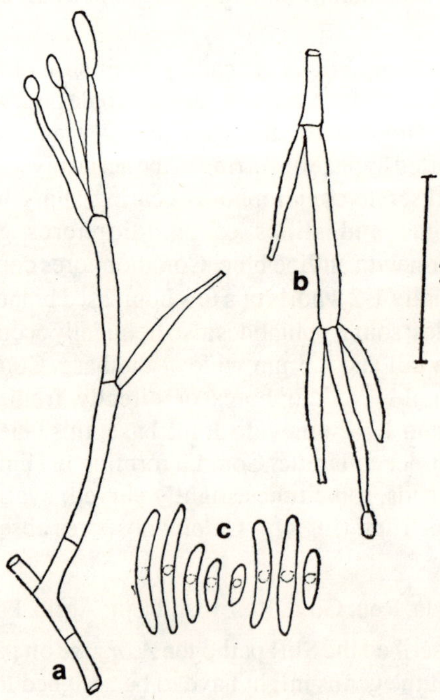

Fig. 5abc : (ab) Conidiophores and phialides from nature and culture respectively, (c) phialoconidia (Agnihothrudu, 1973).

diameter, hyaline, smooth, macronematous, mononematous, once or twice branched, with apical conidiogenous cells in groups of 2-5, monophialidic, discrete, elongate, cylindrical. Single, individual, intercalary phialides also observed on the conidiophores. Phialides flask-shaped, elongate, narrow towards the apex, 10-16 (-18) X1-1.5(-1.5)µm and in culture sometimes as long as 20µm. Phialoconida acrogenous, simple, hyaline, produced successively, cylindrial to sub avicular, 4-10(-12) X 0.5-1 (-1.25) µm in nature, but slightly larger in culture measuring 7-10 (-14) X 0.6-1 (-1.5)µm, generally with a guttule in the middle, as in the case of species of *Colletotrichum* or *Gloeosporium* (Fig. 5abc).

Tasugi and Ikada (1956) studied the conidial variation of *A. oryzae* obtained from host plant (2.1-8.5 x 0.5-1.6 µm) and from culture (1.8-13 x 1-1.6 µm). Further, Ou (1975), Amhead et al. (1975) and Gams and Hawksworth (1975) studied the morphological characteristics of the organisms.

Brady (1980a,b) described *Sarocladium oryzae* and *S. attenuatum* (CMI descriptions of pathogenic fungi and bacteria Nos. 673 and 674) and pointed out that both the species have similar conidiophores. However, ShR pathogen, *S. oryzae* produces less regularly verticillate conidiophores and shorter conidia with rounded ends (3.5-7 x 0.8-1.5 µm) while *S. attenuatum* has more regularly and longer unicellular conidia with truncated ends (4.5 (-13) x 0.6 (-1) µm). Both the species (*S. oryzae* and *S. attenuatum*) co-exist at times and resemble each other except for pigmentation and subtle variation in spore size and shape, the former with round and the later with truncated ends (Gams and Hawksworth, 1975).

Morphology

The morphological features of ShR pathogens described by various workers is summarised (Table 3) by Joe and Manibhushanrao (1996a). Although studies were conducted independently over a period of time, there seems to be not much variation among the features such as vegetative hyphae, phialides and conidia described by several authors.

Species Concept

Bridge et al. (1989) made an extensive study on the revision of the species in the isolates of *Saracladium* spp. from rice and bamboo based on biochemical and morphometric analysis. Fifty isolates of *Sarocladium* spp. were examined for conidial variation (Fig. 6) size and shape (Fig. 7) and physiological characteristics (Table 4) viz., growth, carbon utilization, DNA hydrolysis, esterase isoenzyme bands and production of secondary metabolites. With the exception of the DNA hydrolysis test, there was no appearent correlation of any of the biochemical and physiological alterations and did not produce any characters that could be used to distinguish the two species based on conidial size and shape.

Variation in the metabolites produced by the individual isolates could be distinguished, however, there seems to be several common spots (identified by thin layer chromatography) for isolates named as different species and isolates from different

Table 3: The summary of morphological features of sheath rot pathogens (Joe and Manibhushanrao, 1996a)

Fungus/Author	Range(µm)			
	Veg. hyphae+	Conidiophores	Conidia	Phialides
Acrocylidrium oryzae				
Sawada (1922)	2-2.5	6-20 × 1-1.5	4-9 × 1-2.5	
Tasugi and Ikeda (1956)	-	-	1.8-13 × 1-1.6 2.1-8.5 × 0.5-1.6*	-
Agnihothrudu (1973)	2	4-10(-12) × 0.5-1(-1.25)	(10-16(-18) × 1-15(-1.5)	
Sarocladium oryzae				
Gams and Hawksworth(1975)	1-2.5	60<2 × 2-2.5	3.5-7 × 0.8-1.5	30-40 × 1.2-2
Brady (1980a)	1-1.5	-	3.5-7×0.8-1.5	30-40×1.2-2
Joe and Manibhushanrao (1996a)	1-1.5	-	3-0.08×1×1.8	20-30×1.5-2 16-20 × 1.0-1.5**
Sarocladium attenuatum				
Gams and Hawksworth (1975)	1.5-2.5	60-90 × 2.5	4.5-8×0.6-1	22-45×1.5-2
Brady(1980b)	1-1.5	-	4.5-8(-13) × 0.6(-1)	30-40×1.5-2
Joe and Manibhushanrao (1996a)	1-1.8	-	4.5-12 × 0.8-1.2(-1.5)	30-37.5 ×0.7-1.5

* From host plants; ** Developed on undifferentiated vegetative hyphae, +; dia.

locations and hosts (Fig. 8). This indicated that bamboo blight isolates from Bangladesh are genetically distinct from the isolates from rice in 12 countries. Further, Boa and Brady (1987) and Boa and Rahman (1987) suggested bamboo blight might be transmitted from rice.

Although the species concept has been worked out from both the angles using biochemical and morphological characters there are insufficient grounds to justify the recognition of *S. oryzae* and *S. attenuatum* as distinct species. Hence it can be concluded that *S. attenuatum* is a synonym of *S. oryzae*.

Joe and Manibhushanrao (1996a) described two representative isolates from each *Sarocladium* sp. grown on CDA medium for their hyphal characteristics, conidial initiation and morphology using light and scanning electron microscopic (SEM) studies. The colony and conidial characteristics, and phialide development were given as follows:

In *S. oryzae* slow growth with upper side of pale saffron and a reverse of ochraceous pigmentation was seen, while, fairly fast growth with pinkish bluff and a reverse showing pale salmon pigmentation was seen in *S. attenuatum* isolates on the CDA medium. Besides, conidial shape and size also confirmed the existence of variation in *Sarocladium* spp.

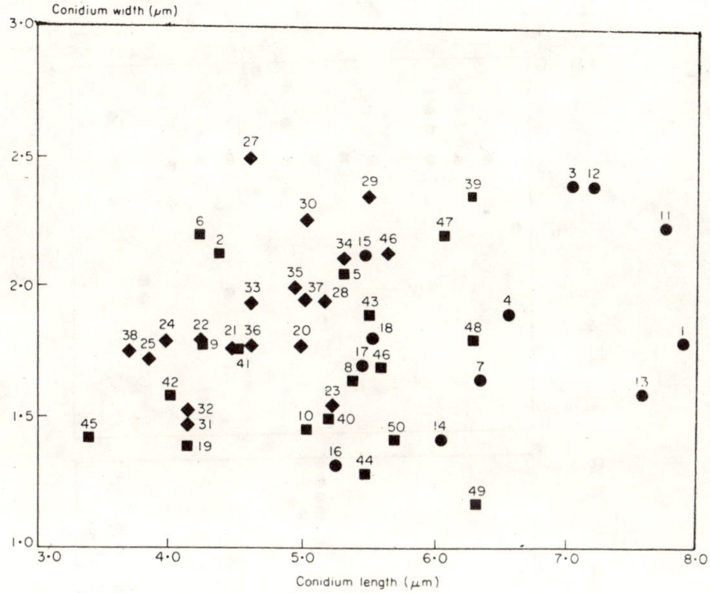

Fig. 6 : Graph of length vs. width for mean conidial measurements of all *Sarocladium* isolates. ●, isolates received as *S. attenuatum*; ■, isolates received as *S. oryzae*; ♦, isolates received as *Sarocladium* sp. Standard deviations varied between 0.36 and 1.56 μm for lengths, and between 0.17 and 0.70 μm for widths (Bridge et al. 1989).

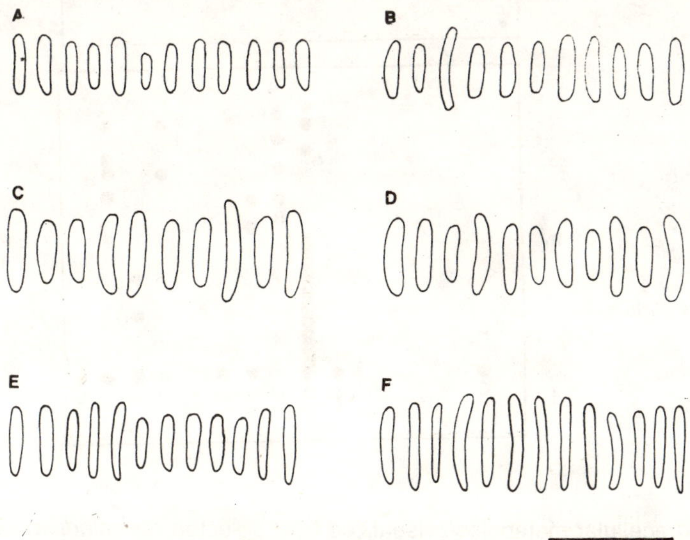

Fig. 7 : Conidia from six isolates of *Sarocladium* to illustrate the variation of conidial shape and size both within and between isolates. The isolates were selected from different parts of the variation range A, No.25 (IMI 287404); B, No.42 (IMI 297168); C, No.46 (IMI 321278); D, No.9 (TAI, holotype of *S. oryzae*); E, No.7 (IMI 184922, isotype of *S. attenuatum*); F, No.11 (IMI 210411). Scale = 10 μm (Bridge et al. 1989).

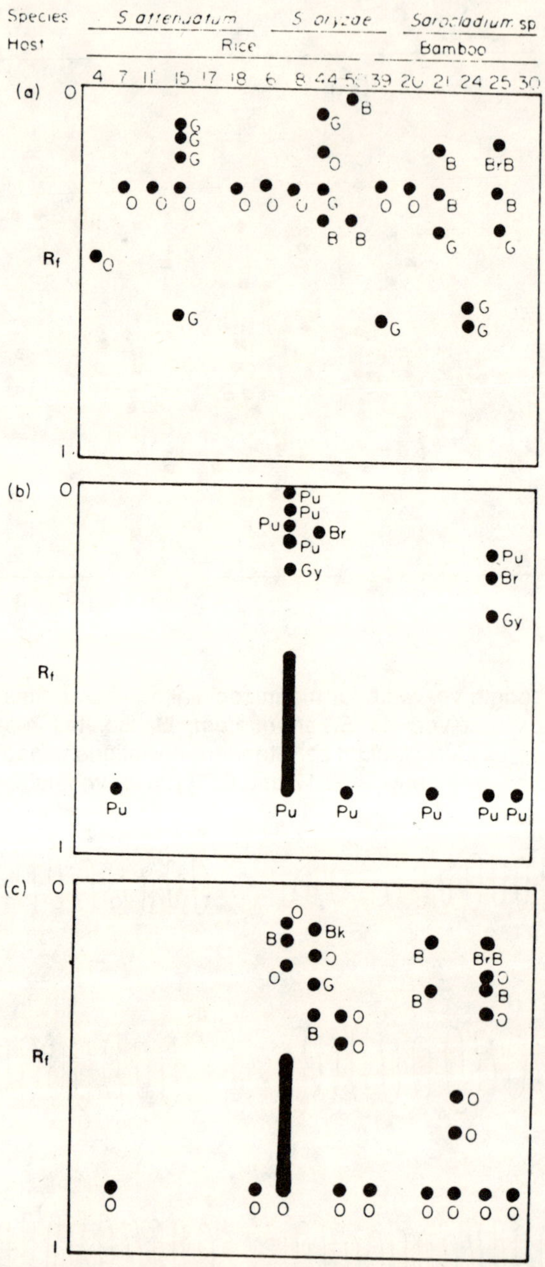

Fig. 8 : Extracellular metabolites visualized from selected *Sarocladium* isolates by thin-layer chromatography. (a) Spots seen under long-wave UV; (b) spots seen under visible light after staining with 0.5% (v/v) anisaldehyde; (c) spots seen under long-wave UV light after staining with 0.5% (v/v) anisaldehyde. B, blue; Bk, black; Br. brown; BrB, bright blue; G, green; Gy, grey; O, orange; Pu, purple. (Bridge et al. 1989).

A *Sarocladium oryzae*

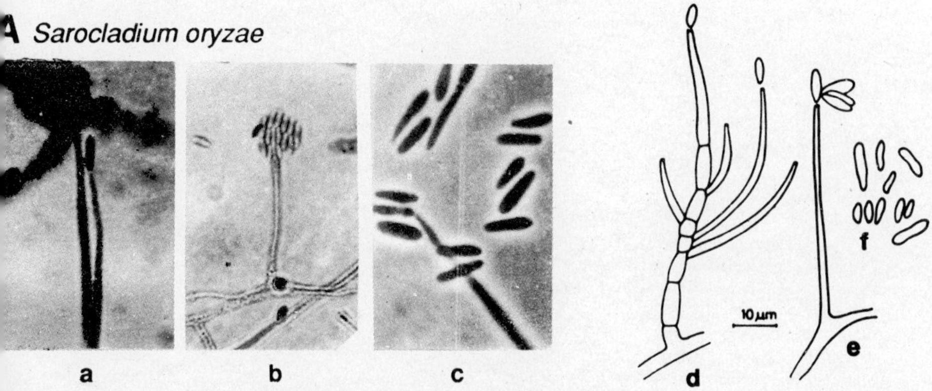

B *S. attenuatum*

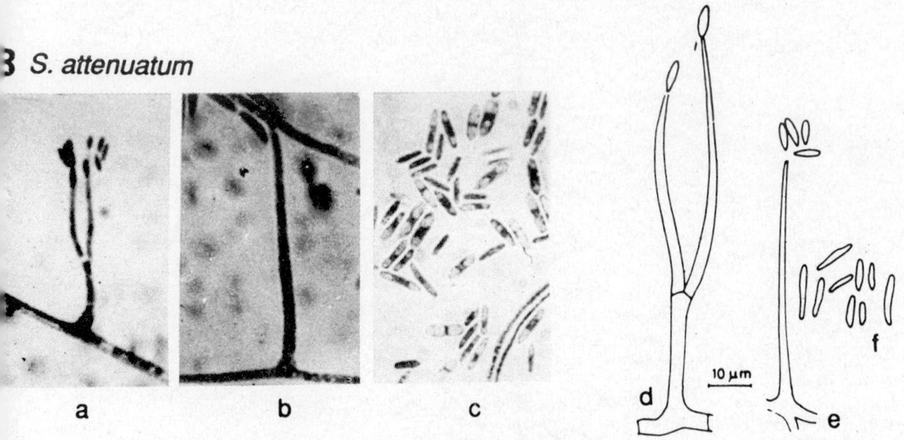

Plate 3AB(a-f). Light micrographs of *Sarocladium* spp. (A) *S. oryzae*, a) Conidiophore with young conidia × 1200, (b) Simple phialide bearing slimy conidia × 360, (c) Conidia × 1350 Camera lucida diagrams: (d) Conidiophores with phialides, (e) Simple phialide, (f) Conidia (B) *S. attenuatum*, (a) Conidiophore with young conidia × 360 (b) Simple phialide bearing slimy conidia × 1200, (c) Conidia × 1200 Camera lucida diagrams: (d) Conidiophores with phialides, (e) Simple phialide, (f) Conidia (Joe and Manibhushanrao, 1996a).

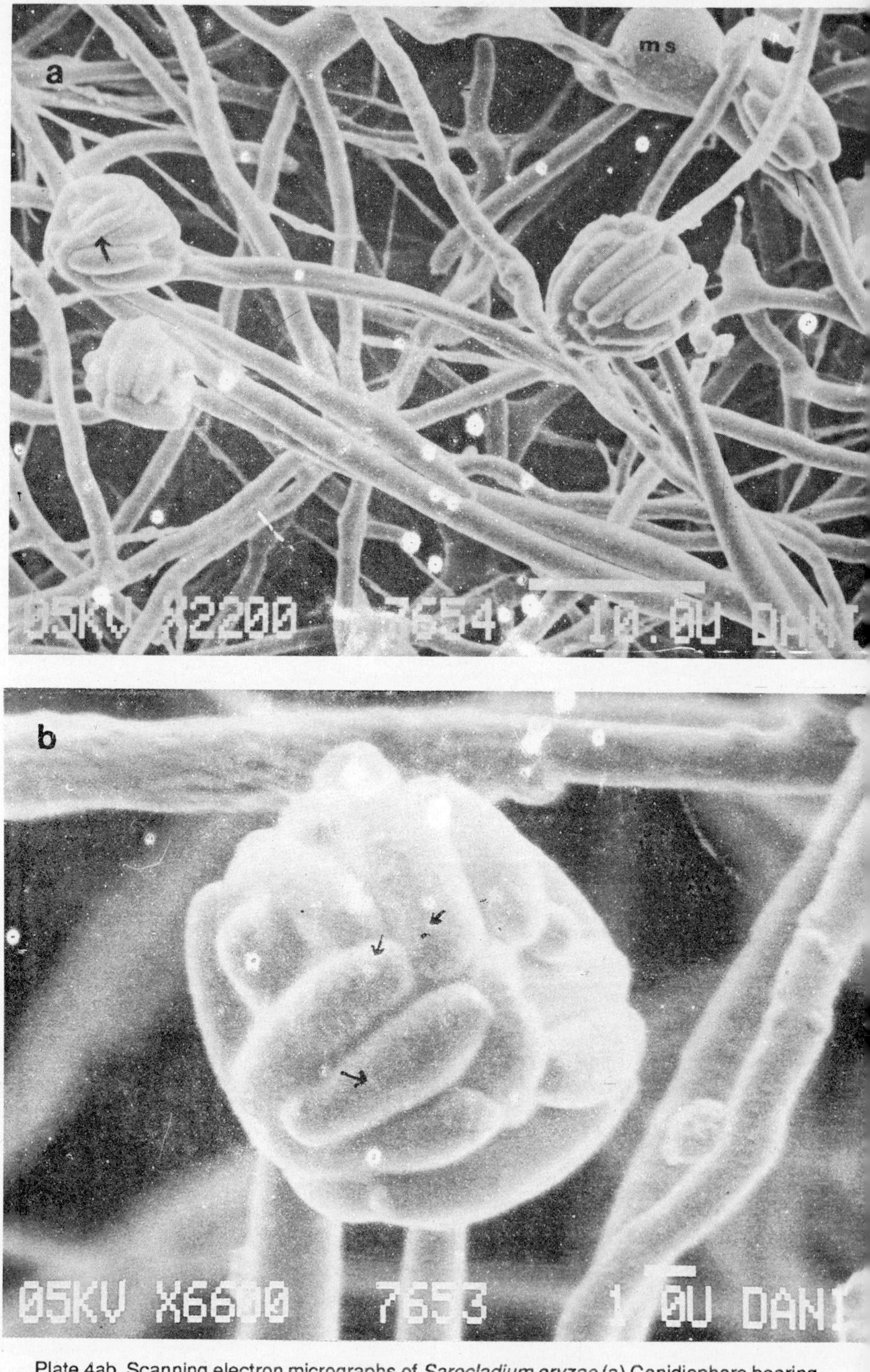

Plate 4ab. Scanning electron micrographs of *Sarocladium oryzae* (a) Conidiophore bearing slimy mass with mucliagenous sheath (ms) and matured conidia seen without ms. Arrow indicates the interwoven hyphae on the slimy head. bar=10µm.
(b) A magnified view of the conidia showing round or blunt ends, arrows indicate the protuberances on the surface of conidia. bar=1µm (Joe and Manibhushanrao, 1996a).

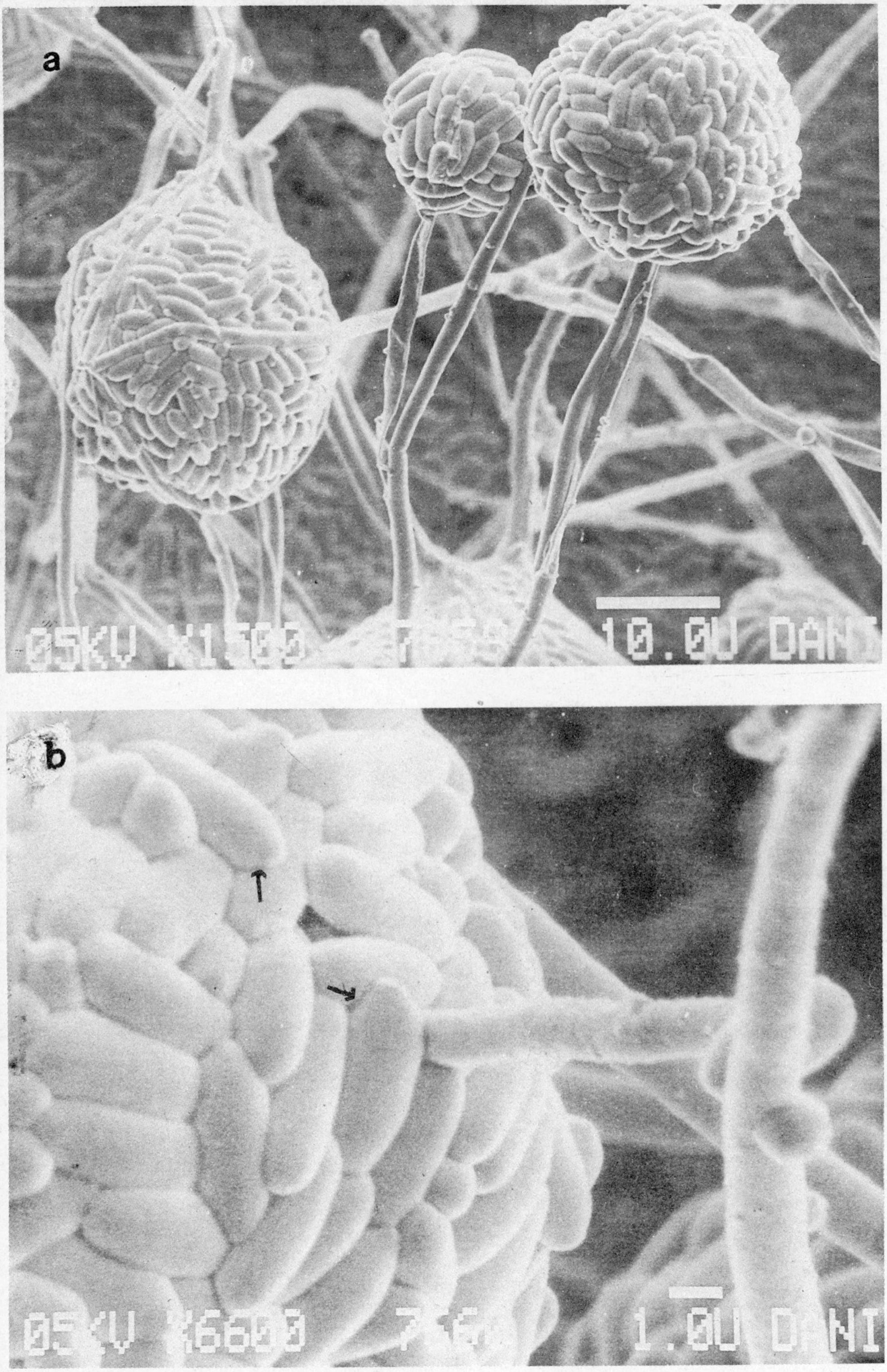

Plate 5ab. Scanning electron micrographs of *Sarocladium attenuatum*. (a) The slender conidiogenous cells bearing mass of conidia. bar=10μm
(b) A magnified view of smooth walled conidia. Arrows indicate truncated ends, bar=1μm (Joe and Manibhushanrao, 1996a).

Table 4: Physiological characterization of isolates (Bridge et al. 1989)

Study number	Growth at 37°C	Lactic acid	Citric acid	Hydrolysis of DNA	Esterase bands	Species	Host
4	+	w	w	+	C D	*S. attenuatum*	Rice
6	+	+	+	+	B F	*S. oryzae*	Rice
7	+	–	+	+	No bands	*S. attenuatum*	Rice
18	+	+	w	+	No bands	*S. attenuatum*	Rice
31	+	w	w	–	D F	*Sarocladium sp.*	Bamboo
32	+	w	w	–	G	*Sacrocladium sp.*	Bamboo
33	+	w	+	–	C E	*Sarocladium sp.*	Bamboo
41	+	w	w	–	D E	*S, oryzae*	Bamboo
42	+	+	_	–	D	*S. oryzae*	Bamboo
4	+	–	+	+	B	*S. oryzae*	Rice
44	+	w	–	+	A B C	*S. oryzae*	Rice
45	–	–	w	+	B	*S. oryzae*	Rice
50	+	–	w	+	B	*S. oryzae*	Rice

In addition, the above isolates, all hydrolysed casein and Tween 80 after 7 d and gelatin after 14 d, used lactose as sole carbon source, failed to grow with ammonium tartrate or tannic acid as sole carbon sources, and failed to grow at 4 and 44°C.

w, weak or inconclusive result.

Easterase band R_f values : A, 0.15-0.20; B, 0.33-0.38; C, 0.58-0.63; D, 0.66-0.71; E, 0.75-0.78; F, 0.83-0.88; G, 0.90-0.95.

Light microsopic studies

Sarocladium oryzae: Colonies on CDA medium reached 5.0 to 7.5 cm dia. in 10 d at 30 ± 2°C, appeared pale saffron, reverse ochraceous (Plate 3A). Spore suspension inoculated plate showed an uniform growth. The average diameter of the vegetative hyphae measures 1-1.8 µm wide. This is in agreement with observations made by the most workers (Table 2). However, Sawada (1922) and Agnihothrudu (1973) reported that the average dia. hyphae ranges from 2-2.5 µm while, Gams and Hawksworth (1975) reported the range as 1-2.5 µm. The phialides arise either on the conidiophores or directly on vegetative hyphae, while the former measured 20-30 µm x 1.5-2 µm in size and the later were slightly smaller (16-20 µm x 1-1.5 µm). Conidia were formed singly or as slimy masses. They were cylindrical with rounded ends, hyaline, thin walled, smooth and one celled (3-8 x 1-1.8-2) µm (Plate 3A (a-d)) (Joe and Manibhushanrao, 1996a).

S. attenuatum: Cultures on CDA medium reached 8 to 9 cm dia. in 10 d at 30 ± 2°C, colonies appeared pinkish-buff and reverse pale salmon. Spore suspension inoculated cultures grew uniformly over the plate in 5 d. Vegetative hyphae were 1-1.8 µm wide. Conidiogenous cell phialidic, originate either as ultimate branch of conidiophore or directly from the undifferentiated hyphae. The phialides were 30-37.5 µm long and 0.7-1.5 µm wide.

Conidia borne singly or seen collectively at the tip, were hyaline, one celled with truncated ends (4.5-12 (-15) x 0.8-1.2(-1.5) µm).(Plate 3B) (Joe and Manibhushanrao, 1996a).

No correlation was obtained between cultural characteristics and virulence of the isolates, in confirmation of the earlier studies on ShR pathogen by Shahjahan et al. (1986).

SEM studies

Unlike the morphological differences (Joe and Manibhushanrao, 1996a), the SEM studies revealed a clear demarcation between the two species especially in the shape, size and surface of the conidia and conidiophore (Plate 4ab and 5ab). Between *Sarocladium* spp., despite their morphological differentiation, they produced identical symptoms on rice sheaths. Bridge et al. (1989) considered *S. attenuatum* as the synonym of *S. oryzae* through morphometric and biochemical analysis of the isolates. On the contrary SEM studies gave distinct differences than similarities to delineate *S. oryzae* from that of *S. attenuatum*.

Among the similarities, the conidia of *Sarocladium* spp. borne directly from the conidiophores aggregate as a globose head and covered in a mucilaginous layer. The mucilaginous sheath appears to disintegrate as the conidia mature and become visible. Some of the globose conidial heads showed interwoven hyphae (Joe and Manibhushanrao, 1996a). The major ultrastructural differences are listed below:

S. oryzae (Plate 4a,b)	*S. attenuatum* (Plate 5a,b)
1. Conidiogenous hyphae were broader than vegetative hyphae	No such demarcation was seen
2. Phialides were broader at the base and tapering towards the apex with smaller globose heads	Phialides were cylindrical, slender with larger globose heads
3. Conidia were cylindrical with round or blunt ends and minute protuberances on the surface of the wall	Conidia were cylindrical with truncated to obtuse ends and smooth wall surface

Information on infection process of ShR pathogens in leaf sheaths of rice plants using electron microscope (EM) is lacking. The EM studies are quite essential to understand clearly the ShR pathogen invasion in rice sheaths. Detectable ultrastructural changes in the host cells is also lacking as this help in understanding the cytological changes to support the biochemical evidence.

Physiology

The fungus *S. oryzae* is nutritionally fastidious and its inability to show good growth and sporulation in culture is well recognised. The growth of the fungus is optimum at 30-31°C, little at 37°C, nil at 13°C and a minimum exposure at 50°C is lethal to the pathogen (Kawamura, 1940). Tasugi and Ikeda (1956) reported a temperature and pH range of 20-28°C and 6.4 respectively for the optimal growth of *S. oryzae* and also the inhibitory effect of certain chemicals, viz., $HgCl_2$ (0.33%), Uspalum (0.1-5%), $CuSO_4$ (0.2-5%), formalin

(1-5%), on the growth of the fungus. Further, Ramakrishnan (1971), Shahjahan et al. (1977), Chien and Thseng (1981) and Parvathi (1979) have studied the nutritional physiology of the fungus and reported suitable culture media and optimal temperature and pH ranges.

Studies on six isolates of the fungus grown on potato dextrose agar, oatmeal agar, Czapek's-dox (CD) agar and CD agar with yeast extract (0.1%) at 25°C and 32°C, showed differential response in their growth pattern and sporulation (Parvathi, 1979). It is pertinent to note here that Chen (1957) also reported such behavioral differences among *S. oryzae* isolates to physiological factors like temperature, pH, and various carbon and nitrogen sources. The pathogen, *A. oryzae* is slow growing in culture (Shahjahan, 1977). However, Parvathi (1979) reported significant growth of the fungus in CD liquid medium supplemented with yeast extract (0.1%) even on 3 d incubation while there was sudden spurt in growth from 4 d onwards till 8 d. A six-fold increase in the biomass production was reported followed by a decline till 16 d. The factor(s) present in yeast extract was not identified despite biochemical tests. It is heat stable and water soluble. An Rf value of 0.1-0.4 in butanol:acetic acid : water, and an absorption peak at 260 nm enhanced the growth of *S. oryzae* (Parvathi, 1979). However, the growth of the fungus was not stimulated by various compounds viz., B-group vitamins, amino acids, nuclic acids, fatty acids, steroids and auxins.

Carbon and nitrogen utilization

The carbohydrate and nitrate utilization on the growth and sporulation is very important in the physiology of the ShR pathogen. Growth and sporulation appear to be dependent on the nature of nitrogen along with the carbon sources and other environmental factors. Earlier, Kawamura (1940) reported that the fungus is fastidious in its nutrient uptake. The growth of *S. oryzae in vivo* was proportional to N content (Hashioka and Makino, 1956). Sucrose was best utilized among various carbon sources (Parvathi, 1979; Singh, 1982). Parvathi (1979) reported the progressive increase in the growth of ShR pathogen in the basal SN (sucrose nitrate) medium where 5 to 30 g sucrose was incorporated and at higher concentration (30 g/l) it was relatively well utilized (140 mg dry wt./50ml medium) over a prolonged period of 18 d incubation (Table 5). However, with SN medium (sucrose 5g/l) optimal growth was obtained on 9 d incubation itself (27 mg dry wt./50 ml medium) and from 12 d onwards autolysis took place. However, in the case of $NaNO_3$ utilization there was a decline on 12 d in all the four concentrations tested. The information on nutritional metabolism of *Sarocladium* spp. is limited. Further, Joe and Manibhushanrao (1995) studied the utilization of different carbon, nitrogen sources with respect to the growth of the differentially virulent isolates of *Sarocladium* spp. (Table 6 and 7).

The variation in carbon utilization between the isolates of *S. oryzae* and *S. attenuatum* was highly significant (P<0.001). Among the carbon sources tested starch was best utilized by the SO1 and SA 1 yielding 15, 7 and 14, 4 g/l respectively (Table 5). CMC and chitin were less preferred by all the 4 isolates tested.

Among the sources tried KNO_3 was the best source for growth of SO 5 and SA 1, while isolates SO 1 and SA 5 utilized it poorly. On the other hand, KNO_3 and glutamic acid were equally was the best sources for isolates SO 1 and SA 5 utilized poorly. The ammonical N of $(NH_4)_2$ NO_3 was the best source for the isolate SO 1 and SA 5 utilized very poorly. However, glycine and urea were equally preferred by all the isolates (Table 7).

Table 5 : Effect of increased concentrations of sucrose on the growth of *A. oryzae* (Parvathi, 1979).

Sucrose g/l	Mycelial dry wt (mg)					
	Days of incubation					
(Basal medium)	3	6	9	12	15	18
5	6	20	27	21	15	12
10	6	23	41	52	46	40
15	5	26	45	55	65	75
20	9	32	48	73	81	110
25	8	49	52	58	95	120
30	8	39	77	85	100	140

Table 6: Growth* of *Sarocladium* spp. in different carbon sources (Joe and Manibhushanrao, 1995)

Isolate	Glucose	Sucrose	Starch	Pectin	CMC
S. oryzae					
SO 1	13.7a	13.0a	15.7a	4.9b	4.1a
SO 5	9.5c	9.8c	8.2c	5.7a	1.3c
S. attenuatum .					
SA 1	10.6b	11.2b	14.4b	4.5b	2.9b
SA 5	4.4d	6.6d	7.0d	5.9a	1.1c

* g dry wt^{-1}

Means within a column sharing different letters differ significantly (P<0.05) by DMRT

Table 7: Growth* of *Sarocladium* spp. in different nitrogen sources (Joe and Manibhushanrao, 1995)

Isolate	KNO_3	NH_4NO_3	Glutamic acid	Glycine	Urea
S. oryzae					
SO 1	9.3b	11.4a	8.6b	9.2a	9.0b
	(8.5)	(7.7)	(8.2)	(7.9)	(6.7)
SO 5	7.8c	7.7b	4.0c	4.0c	4.8d
	(8.4)	(7.2)	(8.4)	(8.1)	(6.8)
S. attenuatum					
SA 1	12.5a	10.2a	10.4a	9.1a	11.5a
	(8.8)	(7.7)	(8.3)	(7.2)	(6.9)
SA	5 6.5c	4.9c	4.6c	6.8b	7.4c
	(8.6)	(7.5)	(7.4)	(7.8)	(7.1)

* g dry wt^{-1} ; Value in parenthesis denotes final pH

Means within a column sharing different letters differ significantly (P<0.05) by DMRT

Sucrose and starch were found to be the best among various carbon sources tried while NH_4 NO_3 and $(NH_4)_2$ SO_4 promoted better growth of the ShR pathogen (Mohan and Subramaniam, 1979b).

Sporulation

The failure to obtain good sporulation and growth in culture of *A. oryzae* is well recognized. Addition of natural products to the CD agar medium did not favour sporulation. However, exposure to near UV light for 4 d stimulated sporulation (7 fold increase) of ShR pathogen over incandesent light (Parvathi, 1979).

Joe and Manibhushanrao (1996a) reported sporulation of the isolates of 13 *Sarocladium* spp. in CDA medium incorporating various carbon and nitrogen sources used in *in vitro* growth studies. The carbon sources which favoured regulative growth also stimulated sporulation. Variation within carbon source/isolate, incubation days differ significantly ($P<0.01$). The monosaccharide, glucose favoured good sporulation but disaccharide sucrose reduced maximum sporulation in isolates SO 1, while maltose supported moderate sporulation of all the isolates. Polysaccharides starch and pectin enhanced sporulation but trisaccharide raffinose found to be a very poor source for sporulation of all the isolates.

Various nitrogen sources used for growth studies were used to study sporulation using CDA medium. All nitrogen sources favoured sporulation, however, significant difference in sporulation was recorded between the N sources/isolates at 0.1% level (Table 8). The most favourable N source for optimum sporulation was found to be urea for all the four isolates. Further, KNO_3 favoured for isolates SA 1 and SO 1, while NH_4NO_3 for SA 1 and glutamic acid for isolate SO 1 for sporulation.

Table 8: Effect of oxidized products of chlorogenic acid on PG activity of *A. oryzae* (Manibhushanrao and Sreenivasaprasad, 1985).

Concentration in (μm/ml)	Enzyme activity(units/ml)		Inhibited activity (units/ml)	Percentage inhibition
	Phenol+ phenolase	Original*		
0.5	16.60		11.9	41.8
1.0	13.80		14.7	51.5
		28.5		
1.5	13.20		15.3	53.7
2.0	15.10		13.4	47.0

*PG activity of the culture of *A. oryzae*.

Toxins

The ShR pathogen *S. oryzae* produces toxic metabolites helvolic acid (Dachene et al. 1968) and cerulenin (Arison and Omura 1971, Omura, 1976; Turner and Aldridge, 1983 and

Berdy, 1986). Cerulenin (helicocerin) is an unsaturated amide, the non-steroid antibiotic and has a molecular formula : $C_{12} H_{12} O_3N$ with molecular wt. 223.26. The structure of cerulenin was confirmed as: 2*R*, 3*S*-epoxy-4-oxo-7,10-*trans*,-*trans* dodecanoic acid amide (Fig. 9). Biosynthetic studies of cerulenin have revealed that its primary skeleton is constructed from six acetate molecules.

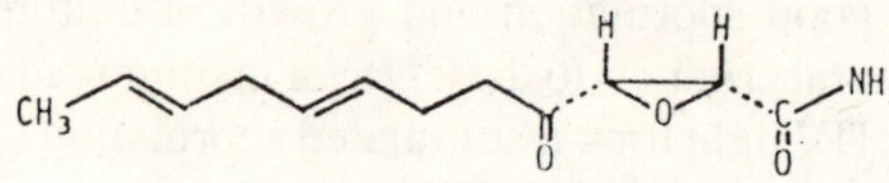

Fig. 9 : Structure of cerulenin (Omura, 1974).

Cerulenin is primarily produced by *Cephalosporium caerulens* and by a few other fungal pathogens including *S. oryzae* (Turner and Aldridge, 1983). Its minimum inhibitory concentration (MIC) for rice blast pathogen *P. oryzae* and ShR pathogen *A. oryzae* are 12.5 and 6.75 µg/ml respectively.

Bioassay for cerulenin by filter paper disk method using *Candida albicans* (KF-1 strain) was reported by Omura (1981). *In vitro* cerulenin production by the ShR pathogen (Iwai et al. 1973) as well as *in vivo* detection ShR infected rice tissue (Sakthivel, 1987) were available. Further, the phytotoxicity of cerulenin on several rice cvs as well as two weed hosts was done by Sakthivel (1987) and the symptom developed were similar to that of ShR infection. He also studied the antimicrobial activity especially towards pathogenic fungi (*Geotrichum candidum* and *Candida albicans*) and bacteria (*Escherichia coli, Bacillus subtilis* and *Pseudomonas fluorescens*).

Gnanamanickam and Mew (1991) reported that the toxin and / or antibiotics produced by *S. oryzae* showed toxicity towards rice pathogens (*Sclerotium oryzae, Gaeumannomyces graminis* var. *graminis, Rhizoctonia solani* and *M. grisea*) similar to that of ShR pathogen itself while, stem rot (StR) pathogen *Sclerotium oryzae* was most sensitive to cerulenin and not toxic to *Fusarium moniliforme*. At 2 µg mL^{-1} cerulenin concentration the diameter of the inhibition zone in *S. oryzae* were 4-5 mm increased to 28, 45, 48 mm at 10, 20, 60 µg mL^{-1} respectively (Plate 6). ShB pathogen, *R. solani* showed inhibition zone of 25 and 28 mm dia. at 60 and 90 µg mL^{-1} cerulenin. However, no conclusive evidence on the involvement of cerulenin on the antagonistic interactions with other pathogens of rice was observed.

Vasantha Devi and Gananamanickam (1993) reported the role of cerulenin with the ShR symptom development. Artifical application of cerulenin induced rotting of rice sheath identical to that of induced by *S. oryzae*. The production of the toxin was assessed by the inhibition of *Candida albicans* KF-1 in PDA plates. Virulent isolates produced large inhibition zone (Plate 7) while a less virulent one produced smaller inhibition zones. The maximum inhibition (30 mm dia.) was obtained by the most virulent isolate No. 22 and least virulent isolate No. 6, a mere 6 mm dia. zone. Potato dextrose agar (PDA) assayed plates with lauric and oleic acids at 1-1000 ppm reversed the action of the toxin. This confirms the presence of cerulenin as it inhibits the fatty acid synthetase.

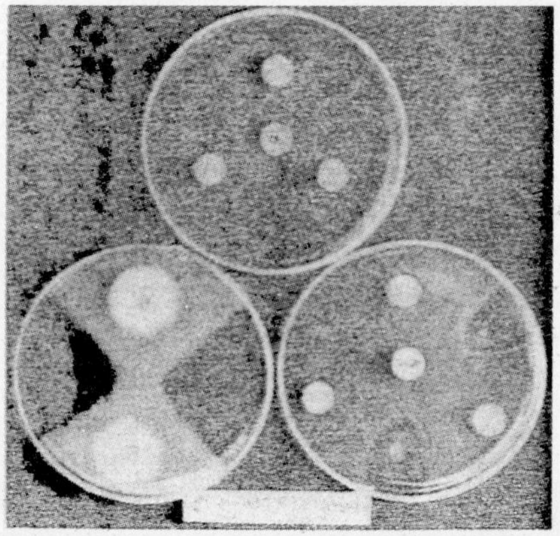

Plate 6. Inhibition of *Sclerotium oryzae,* stem rot pathogen of rice by *Sarocladium oryzae* (seen as white mycelial growth in the plate) and by its toxin, cerulenin (20μg/mL^{-1}; T-20 (plate on the right). Control plate(ck) on top has uninhibited growth of *Sclerotium oryzae* (Gnanamanickam and Mew, 1991)

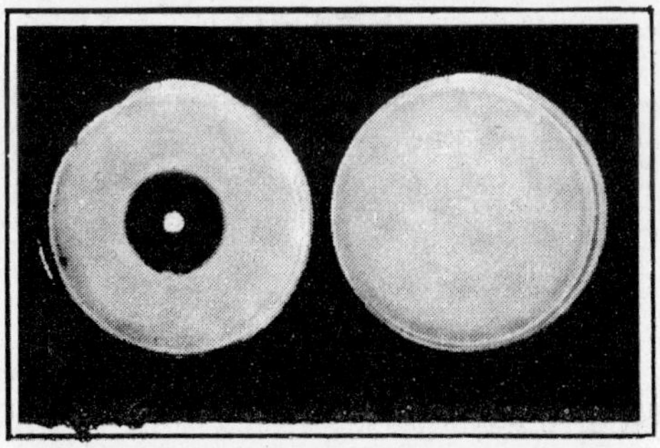

Plate 7. *Candida albicans* KF-1 inhibition assay for cerulenin production of *S.oryzae* and control (Vasantha Devi and Gnanamanickam, 1993).

Plate 8. Induction of ShR symptoms (indicated by arrows) by wild-type (a) and no induction by cerulenin-negative mutant (b) isolates of *S. oryzae*. (Vasantha Devi and Gnanamanickam, 1993).

The UV induced cerulenin-negative mutants isolates No. 6 and 22 did not inhibit *C. albicans* KF-1 or other fungal pathogens of rice including *P. grisea*, *Helminthosporium oryzae*, *R. solani* and *G. graminis* var *oryzae* in plate assay technique while the cerulenin producing wild type isolates did. Further, the mutant isolates did not induce ShR lesions on susceptible rice plants on the other hand, wild type isolate did produce ShR infection (Plate 8). These results further suggest that the cerulenin, the toxin produced by *S. oryzae* may play a role in the pathogenicity of the rice host as well as interaction with fungal pathogens of rice.

Enzymes

Penetration and invasion of host tissue by a plant pathogen is substantially enhanced by the production of cell wall degrading enzymes especially cellulase and pectinases. Pectic enzymes either macerate plant tissue directly or render plant cell walls more susceptible to attack by other cell wall degrading enzymes. While invertase and amylases have a nutritional role which are degradable cellular materials such as sucrose and starch respectively.

The ShR pathogen, *S. oryzae* produces cellulolytic, pectinolytic and oxidative enzymes in culture filtrates (Singh, 1982; Malathy, 1984). However, the production of enzymes were not correlated with virulence of the isolates. Inhibition of polygalacturonase(PG) activity of *S. oryzae* by the oxidized products of chlorogenic acid was reported by Manibhushanrao and Sreenivasaprasad (1985). A maximum of 53.7% repression of the PG activity of *A. oryzae* was effected by the oxidized products of chlorogenic acid at the concentration of 1.5µm/ml of the incubation mixture (Table 9). However, phenol concentrations both above and below 1.5µm/ml showed less inhibition on the PG activity of *A. oryzae*. Here, the inhibition of PG of *A. oryzae* by the oxidized products of chlorogenic acid could either be due to the oxidation of the SH groups of the enzyme by the quinone or by the formation of 1,4 addition compounds by the quinone with the amino groups of the enzyme. Further, critial studies on these aspects would reveal the exact mode of action of the oxidized products of chlorogenic acid on the PG of *A. oryzae*.

Table 9: Sporulation of *Sarocladium* spp. in different nitrogen sources (Joe and Manibhushanrào, 1996a)

Isolate	KNO_3	NH_4NO_3	Glutamic acid	Glycine	Urea
S.oryzae					
SO 1	249a	153b	133a	74b	244a
SO 5	128b	105c	44c	34c	117c
S.attenuatum					
SA 1	124b	178a	130a	100a	196b
SA 5	48c	51d	72b	12d	73d

Means within a column sharing different letters differ significantly (P<0.05) by DMRT

Germination of rice, barley, wheat, rye seeds and the conidia of the rice blast pathogen, *P. oryzae = M. grisea* was inhibited by the culture filtrate of the ShR fungus (Chung, 1975). In our studies (unpublished) also, the culture filtrate (5 fold concn.) of *S. oryzae* induced localized browning reaction on the rice sheath tissue.

Biochemical constituents

Differentially virulent isolates of ShR pathogens (details are given in chapter pathogenicity) *S. oryzae* (SO 1 and SO 5) and *S. attenuatum* (SA 1 to SA 5) were tested for their relative *in vitro* production of various enzymes viz., CMCase, PTE, invertase and protease and identified *S. oryzae* isolates as the most aggressive (Joe and Manibhushanrao, 1995). The biochemical constituents involved showed a positive correlation with the relative degree of virulence of the isolates on rice plants.

The production of various extracellular enzymes by the pathogenic isolates SO 1, SA1, SO 5 and SA 5 was on par with the utilization of various sugars as well as their relative virulence (Fig. 10). However, the highest ß-D- glucosidase activity exhibited by the weakly virulent isolates (SA 5) is not comparable with sugar uptake and the disease index (DI). In general, *S. oryzae* isolates showed more enzyme activity than the *S. attenuatum* isolates indicating the former species had potential factor to invade host cells during colonization.

The growth of isolates of *Sarocladium* spp. was not appreciable in CMC- amended medium which might be due to their lesser ability to produce CMCase. However, this activity could well be correlated with the relative virulence of the isolates (Fig. 10).

Pectin was moderately utilized by all the isolates of *Sarocladium* spp. Of these, the mildly virulent SA 5 utilized the maximum, despite exhibiting the lowest pectin transeleminase (PTE) activity. This showed that the substrate utilization is not in accordance with the enzyme production. No difference in PTE production was observed between mildly virulent isolate of *S. oryzae* (SO 5) and the highly virulent isolate of *S. attenuatum* (SA 1), though the later caused more infection. Such observations do not clearly emphasize the role of PTE. Earlier the *in vitro* production of pectinase by the isolates of ShR pathogen, *A. oryzae* was reported (Kimura and Mizyshima, 1973), however, their role in pathogenesis was not established.

The higer invertase activity of the virulent isolates of *Sarocladium* spp. would enable pathogen colonization in the host tissue. Such an increased expression of invertase and amylase may have a major role in hydrolyzing the respective sugar components, revealing *Sarocladium* spp. as "sugar loving fungi". Carbohydrase and protease activities were more in *S. oryzae* isolates than in *S. attenuatum* and was higher in highly virulent ones. The degradation products *in situ*, the amino acids and amides might serve as major N sources to the ShR pathogen isolates for subsequent colonization and spread. Further, the higher peroxidase (PO) activity explains pathogenic origin and this might be a crucial moiety for the symptom expression.

These results clearly indicate that the virulence of the pathogen is governed by certain biochemical factors which are associated with pathogenesis. The mycelial constituents and enzyme activities relegated to show a relatedness between the isolates of *S. oryzae* and *S. attenuatum* though the former was aggressive in its biochemical determinants as well as

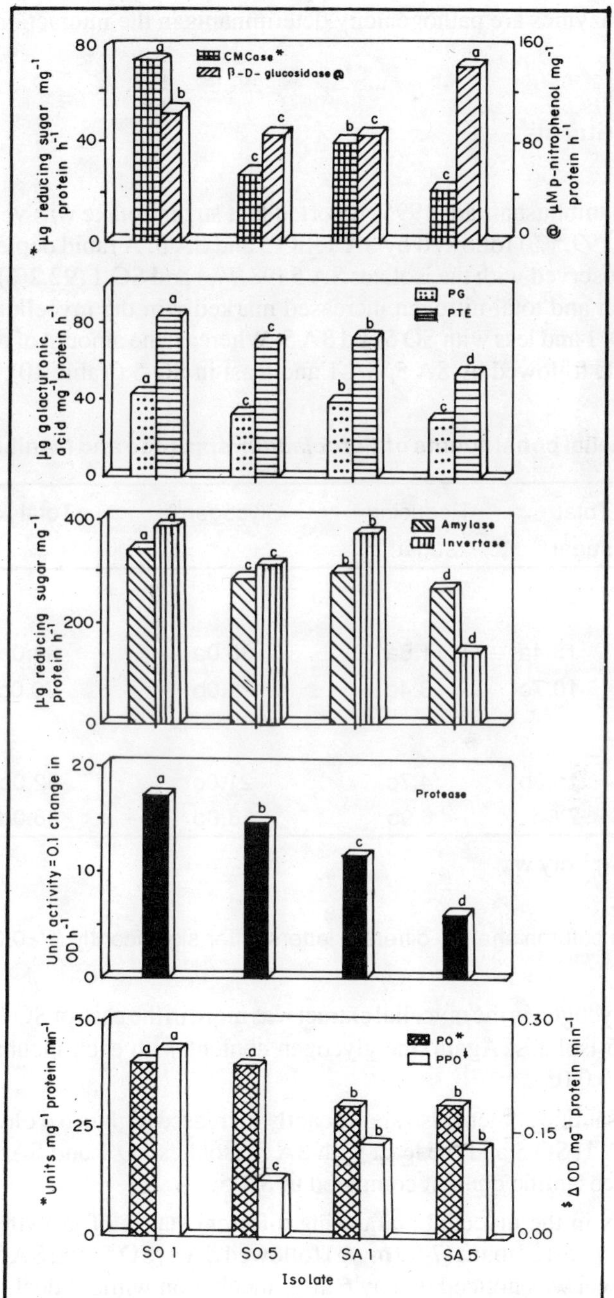

Fig. 10 : Extracellular enzyme production by pathogenic isolates of *Sarocladium oryzae* (SO 1 and SO 5) and *S. attenuatum* (SA 1 and SA 5). Uniform bars in each enzyme shared by different letters differ significantly ($P < 0.05$) by DMRT (Joe and Manibhushanrao, 1995).

disease expression. Further, a molecular approach is needed to determine whether or not the extracellular enzymes are pathogenicity determinants in the interaction of *S. oryzae* with its rice host.

Mycelial constituents

Joe and Manibhushanrao (1995) reported that sugar uptake was very high in the case of isolate SO 1 (99.5%) followed by SA 1, SO 5 and SA 5. A rapid depletion of N from the medium was observed with the isolates SA 5 (98.7%) and SO 1 (92.2%). The total storage sugar, glycogen and total nitrogen increased markedly in the mycelium of isolate SO 1, followed by SA 1 and less with SO 5 and SA 5. Whereas, the amount of reducing sugar was highest in SO 1, followed by SA 5, SA 1 and least in SO 5 (Table 10).

Table 10: Mycelial constituents of *Sarocladium* spp. (Joe and Manibhushanrao, 1995)

Isolate	Total sugar@	Reducing sugar@	Glycogen@	Total N#	Total protein
S. oryzae					
SO 1	18.4a	11.8a	35.0a	94.0a	70.0a
SO 5	10.7c	3.4c	26.0b	70.0c	39.0c
S. attenuatum					
SA 1	15.3b	4.7c	21.0c	82.0b	72.0a
SA 5	7.8d	6.9b	18.0d	68.0d	59.0b

@ mg glucose g^{-1} dry wt.

#mg g^{-1} dry wt.

means within a column sharing different letters differ significantly ($P<0.05$) by DMRT.

Further, glycogen in the mycelial extract was more in the case of SO 1 followed by SO5, SA 1 and SA 5 isolates. Again, the glycogen content with each incubation day differed significantly ($P<0.05$).

The glucosamine content also significantly increased high in 6 d old mycelium of SO1 followed by SA 1, SO 5 and the least with SA 5. Isolates SO 1 and SA 1 showed 2-3 fold increase in glucosamine content compared to SA 5.

The total N in the mycelial mat during different stages of growth showed that SO1 registered the highest amount (7-99 mg/g) followed SA 1, SO 5, and SA 5. In the mycelium maximum N level was noticed on day 6 after incubation while it declined on day 9.

The intracellular protein content of the differentially virulent isolates of *Sarocladium* spp. was higher than the content of extracellular protein. The highest quantitative mycelial protein was recoded with isolates SO 1 with 161 mg/g dry weight followed by SA 1 (114 mg), SO 5 (81 mg) and SA 5 (59 mg).

It is clear from the results obtained by the nutrient utilization of the differentially virulent isolates of *Sarocladium* spp. that there is a positive correlation between virulence and the rate of C/N utilization.

Protein profile

No definite correlation was obtained with relative virulence and protein profile of the mycelial extracts of the four differentially virulent isolates of *Sarocladium* spp. (Joe, 1994; Joe and Manibhushanrao, 1996c). They were analysed using native polyacrylamide gel electrophoresis (PAGE), a technique to distinguish the features between the species. However, within isolates of SO 1 and SO 5 of *S. oryzae* there appears to indicate the nature of differential virulence (Fig. 11abc). Thus, the observed variance in the protein profiles would offer some evidence to compare these isolates of *Sarocladium* spp. Nevertheless, these findings may support the results obtained from the antigenic comparison that these species are closely related.

Serological relationship

There exists a serological relationship between plants and their fungal pathogens (Chakraborthy and Purkayastha, 1983; Charudattan and DeVay, 1970 and Kalyanasundaram et al., 1975) although a few instances where such relationship have not been established (Carrole et al., 1972).

It is well known that antigens on the cell wall are characteristic of the organism's surface, antibodies against these antigens as they may serve as an useful tool to distinguish isolates along with morphological differences. Immunoserological techniques in determining specificity or differential virulence in pathogenic fungi seems to be well progressed. Earlier work with rice blast pathogen *P. oryzae,* virulence and avirulence are reflected in antigenicity (Manibhushanrao et al., 1971). Here the affinities and differences in cultural and morphological characters of differentially virulent isolates of ShR pathogen should be brought out by the antigenic analysis to study serological relationship.

Crossed immunoelectrophoresis

Purkayastha and Ghosal (1985) reported the antigenic relationship between three isolates of ShR pathogen *A. oryzae* and a number of resistant and susceptible cvs of rice using immunoelectrophoretic (IE) and crossed immunoelectrophoretic (CIE) techniques. Antigen of cv. Jaya produced a common band with both heterologus and homologus antisera (Fig. 12ab). One common precipitin band was detected between cvs Mashuri and IR 8, Mashuri and CR-126-42-1, and Jaya and CR-126-42-1, but not between Jaya and Ratna. The precipitin band between the antigen preparation of Jaya and the ShR pathogen isolate P1 was found to be similar but not common precipitin band was observed between cv. Mashuri and isolate P1. No antigenic relationship was observed between two ShR resistant rice cvs and three isolates of *A. oryzae*. The weak cross reaction between antiserum

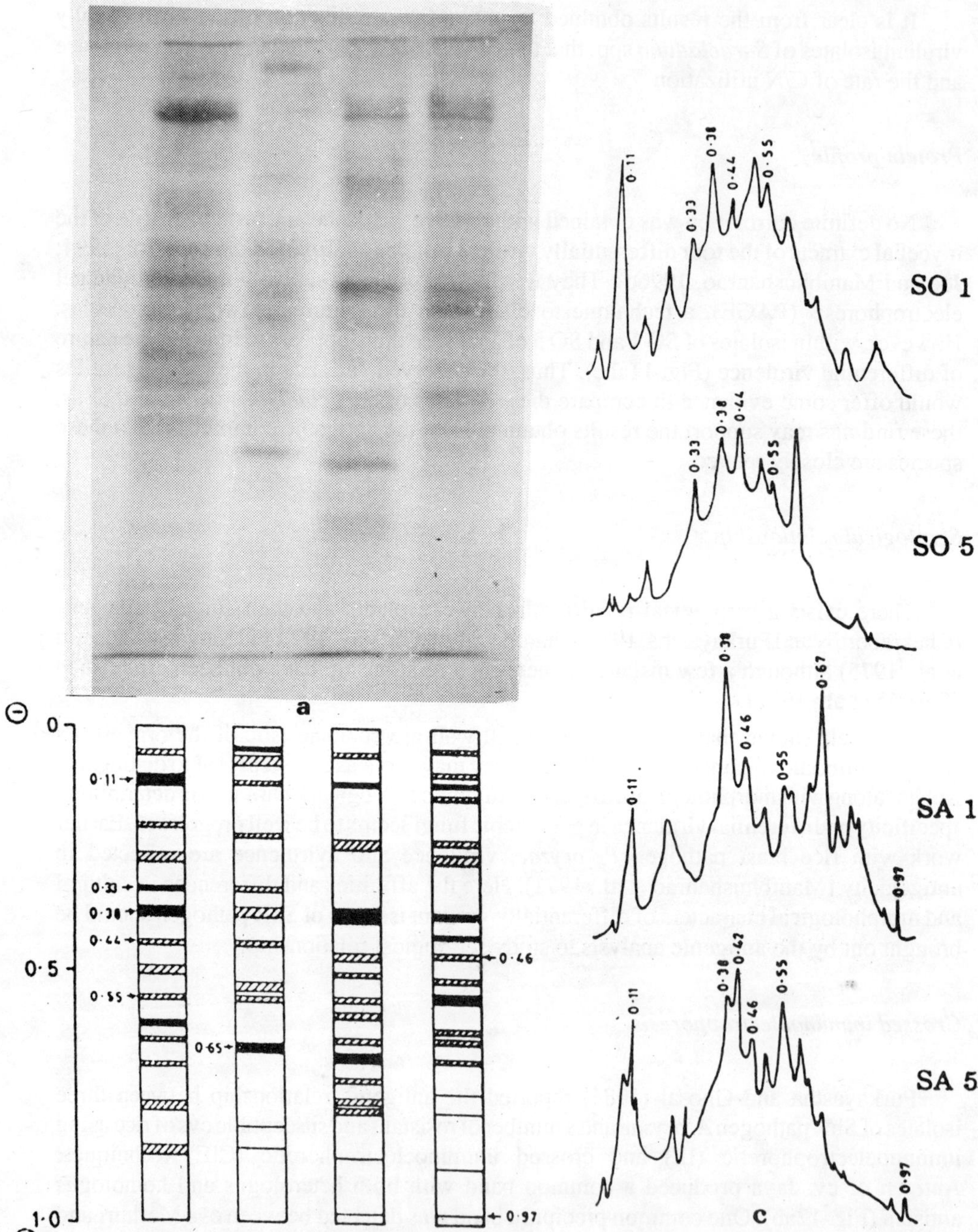

Fig. 11abc : (a) The mycelial protein patterns of *S. oryzae* (SO 1 and SO 5) and *S. attenuatum* (SA 1 and SA 5) (b) composite diagram of protein bands (c) densitometer scans of polyacrylamide gel after electrophoresis (Joe, 1994).

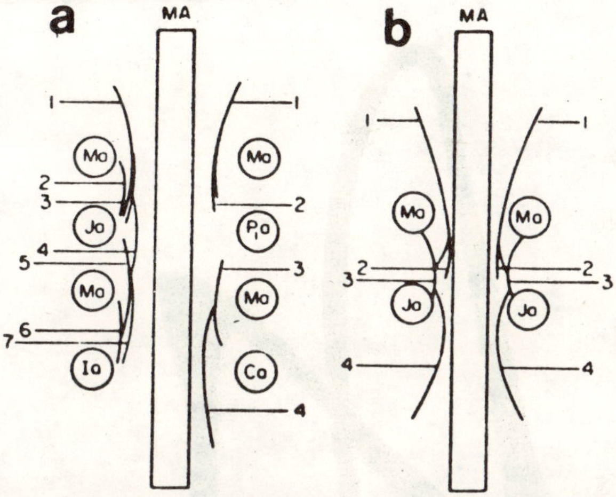

Fig. 12ab : Schematic diagrams of crossed immunoelectrophoretic patterns between antigens of rice and *Acrocylindrium oryzae*. The central rectangular troughs contained antiserum of rice cv. Mashuri (MA) and the surrounding wells contained (a) antigens of cvs Mashuri (Ma), Jaya (Ja), IR-8 (Ia), CR -126-42-1 (Ca) and of the pathogen *Acrocylindrium oryzae* (P_1a), and (b) antigens of cv. Mashuri (Ma) and Jaya (Ja) (Purkayastha and Ghosal, 1985).

of *A. oryzae* (isolate P 1) and antigens from the two susceptible cvs. Ratna and IR-8 may be attributed to low concentrations of antigens or the production of antibodies with low affinity for the antigens. Further, the presence of only one common band between susceptible rice cvs and ShR pathogen *A. oryzae* is a common feature and this was observed with other rice-blast fungal pathogen (*P. oryzae*) combinations (Manibhushanrao et al., 1971). Recently, Joe and Manibhushanrao (1996c) demonstrated the serological relationship among the differentially virulent isolates of *Sarocladium* spp.(Chapter on Pathogenicity) by raising specific antibiodies against the antigens using immunoserodiagnostic techniques.

Antigen analysis showed that when anti *S. oryzae* (SO 1) antiserum was reacted with homologus as well as heterologus *S. attenuatum* (SA 1) antigens, strong precipitin lines were formed. Such a reaction with antigens of these species indicated some serological relationship. However, the appearence of hazy precipitin bands between each of the homologus antigens were analysed by CIE technique. Such precipitin bands common to both *Sarocladium* spp. is suggested of their being the main antigenic constituents of these isolates. The crossed immunoelectrophoretic analysis showed four major antigens (Fig. 13) indicating some serological relationship between these species.

However, cross-absorption test indicated that isolate SO 1 of *S. oryzae* differed from SA 1 of *S. attenuatum* antigenically, indicated the formation of a precipitin band even after absorption tests showed that the SO 2 to SO 5 of *S. oryzae* and SA 3 and SA 4 of *S. attenuatum* were antigenic types of the isolate SO 1. This study suggested that the isolate of the later were antigenically related to the former while differences were noted with SA 2 and SA 5 isolates (Table 11) and hence they are considered serotypes. Thus, different virulent isolates of *Sarocladium* spp. could not be clearly distinguished serologically.

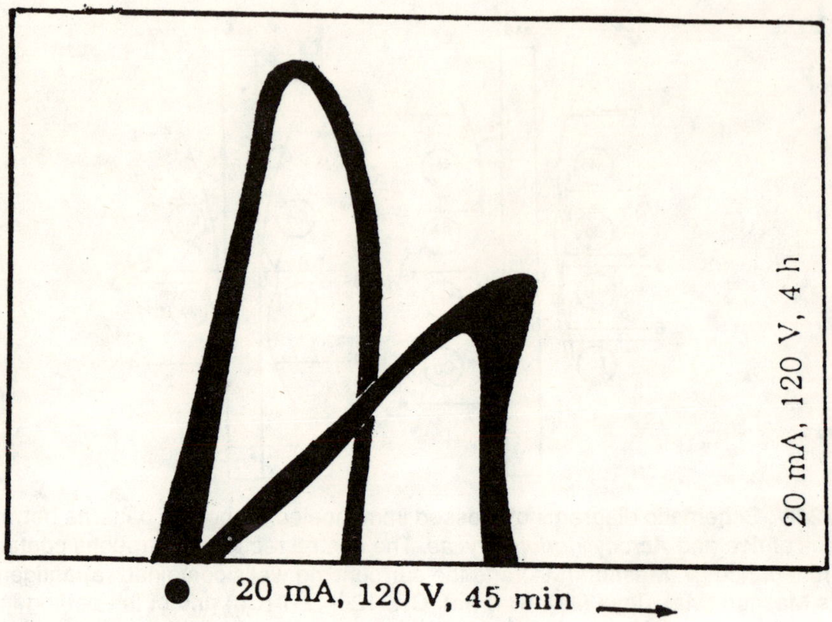

Fig. 13 : Antigenic analysis of *S. oryzae* by crossed immunoelectrophoresis (Joe, 1994).

Table 11: Antigenic typing of isolates of *Sarocladium* spp. by cross absorption (Joe, 1994)

Isolate	Anti SO 1 antibody		Anti SA 1 antibody		Antigenic type
	Unabsorbed	absorbed with SA 1	Unabsorbed	absorbed with SO 1	
S. oryzae					
SO 2	+	+	+	-	SO1
SO 3	+	+	+	-	SO1
SO 4	+	+	+	-	SO1
SO 5	+	+	+	-	SO1
S. attenuatum					
SA 2	+	+	+	+	mixed
SA 3	+	+	+	-	SO1
SA 4	+	+	+	-	SO1
SA 5	+	+	+	-	mixed
SO 1 (control)	+	+	+	-	
SA 1 (control)	+	-	+	+	

+ Precipitin reaction -No reaction

Although conflicting reports on the generic and species of ShR pathogen are available (Agnihothrudu, 1973 and Gams and Hawksworth, 1975) the isolates of *Sarocladium* spp. were morphologically distinguishable (Chapter on Causal organism). However, the apparent presence of common antigens probably justify them to be considered as a single species. Further, critical studies on biochemical markers along with morphological characters are quite necessary for drawing definite conclusions.

The morphological differentiation was envisaged through light and SEM studies. However, the ShR symptoms caused by these isolates of both the species were indistinguishable on rice.

Pathogenicity

There have been conflicting reports on the influence of host nutritional physiology on disease development. Hashioka and Makino (1956) reported the correlation between the host N nutrition and susceptibility to ShR disease. Tasugi and Ikeda (1956) recorded a decrease in ShR incidence at increased N application. Further, the increase and decrease in the ShR disease incidence at higher levels of N and K, respectively, were reported (Chen and Chien, 1964 and Chien and Chu, 1970). However, other investigations (Srinivasan, 1981; Muralidharan and Venkatarao, 1980 and Akanda et al. 1984) revealed a marked increase in the host susceptibility to ShR with increasing doses of N nutrition.

Mode of infection

Little information is available on the mode of infection of ShR pathogen to rice host. The pathogen enters either through stomata or injuries caused by insects and colonize the host vascular bundles and mesophyll tissue intercellularly (Shahajan et al. 1977). Further, histopathological studies are necessary to understand the exact mode of penetration and invasion of the host tissue. Mukherjee and Yadav (1989) made an attempt to study the mode of infection of ShR pathogen *S. attenuatum* on rice. However, these results are inconclusive.

Virulence of the isolates

Use of standard methods is a very important criterion to test the virulence of the isolates of *S. oryzae*. Based on the type of lesions developed on susceptible cv. CO 43. Joe and Manibhushanrao (1995) have studied the virulence of the isolates of both the pathogens and ranked as per the disease index (DI).

Among the several *S. oryzae* and *S. attenuatum* isolates tested, one which caused the maximum infection (DI 88.3) was designated as SO 1 and the least (DI 50.2) as SO 5, while the intermediate ones were ranked from SO 2 to SO 4. Similarly, in the case of *S. attenuatum*, isolates DI from 71.4 to 30.4 were ranked as SA 1 to SA 5 in order (Table 12). The rice plants infected by the virulent isolate (SO 1) had greater numbers of lesions throughout the leaf sheaths and even suppressed the panicle emergence. While infection due to the weakly virulent isolates (SO 5 and SA 5) exhibited only a few lesions on a confined area of the leaf sheath (Plate 9a-e). Generally, *S. oryzae* isolates showed more virulence compared to *S. attenuatum* isolates (Joe and Manibhushanrao , 1995).

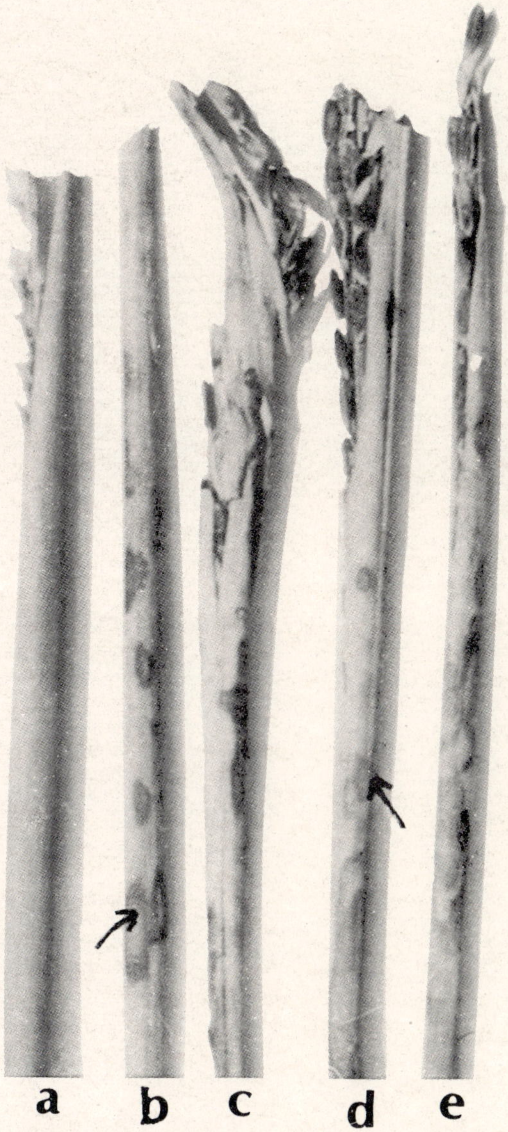

Plate 9a-e Differential reaction on leaf sheaths of susceptible rice cv. CO 43 to highly virulent and mildly virulent isolates of *Sarocladium oryzae* (SO 1 and SO 5) and *S. attenuatum* (SA 1 and SA 5). (a) Healthy leaf sheath, (b) SO 1, (c) SO 5, (d) SA 1 and (e) SA1 and SA 5 isolates (Joe and Manibhushanrao, 1996a).

Table 12: Relative virulence of isolates of *Sarocladium* spp.(Joe and Manibhushanrao, 1995).

Isolate	DI*	Virulence ranking
S. oryzae		
SO 1	88.3a	I
SO 2	70.5b	II
SO 3	68.3c	III
SO 4	62.9d	IV
SO 5	50.2f	V
S. attenuatum		
SA 1	71.4b	I
SA 2	63.1d	II
SA 3	55.4e	III
SA 4	43.5e	IV
SO 5	30.4h	V

*Disease incidence on susceptible rice cv. CO 43

Means of DI sharing different letters differ significantly ($P<0.05$) by DMRT.

Host Range

Balakrishnan and Nair (1981) recorded *Echinochloa crusgalli* (L.) Beauv., *Eleusine indica* (L.) Gaertn., *Monochoria vaginalis* (BURM.f.) Presl., *Cyperus teneriffae* Poir, *C. iria* L., and *C. difformis* L. for the first time as weed hosts of the ShR pathogen *A. oryzae*. Subsequently among the 20 species of weeds and wild rices tested for ShR disease 13 (*Digitaria ciliaris*, *Echinochloa colona*(L.) link;, *E. crus-galli* var., *hispidula*, (Retz.) Honda *E. gabresens*, *O. sativa* var. *rufipong*, *E. indica*, *Paspalidium flavidum*, *Chloris barbata* Sw., *Ischaenum rugosum* Salisb. and *Leptochloa chinensis*(L.) Gerten. were infected by the fungus (IRRI, 1981). However, the rice field weed, *E. colona* was identified as an alternate host of *S. oryzae* (Rahman et al. 1982). The three grasses viz., *E. colona, E. crus-galli* and *C. difformis* with typical symptoms were found in and around ShR disease affected rice fields. The pathogen also causes blight in *Bambusa* sp. (Boa and Brady, 1987).

Deka and Phookan (1992) carried out pathogenicity tests on 10 weeds (Table 13) using an isolate of *S. oryzae* commonly found near rice fields. Only three weeds did not exhibit any susceptible reaction while all others produced typical ShR symptoms. A cross-inoculation test among the seven susceptible weeds with the pathogen isolate was also successful with the symptoms being similar to those that develop when inoculated with rice isolates. The identity of the pathogen was also confirmed as *S. oryzae*. These results also confirm the identity of three new weed hosts (*H. assamica*, *L. hexandra P. walense*) of the pathogen.

Table 13: Pathogenicity of *Sarocladium oryzae* on some weeds commonly found near rice fields. (Deka and Phookan 1992)

Weed	Reaction	Virulance of initial symptom (h after inoculation)
Echinochloa colona (L.) Link	+	72
Monochoria vaginalis (Burm.f.) Presl.	+	120
Hymenachne assamica (Hook f.) Hitchc.	+	168
Leersia hexandra Swartz.	+	144
Panicum walense Mez.	+	144
Oryza rufipogon Griff.	+	72
Eleusine indica (L.) Gaertn.	+	96
Axonopus compressus (Swartz.) Beauv.		
Cynodon dactylon (L.) Pers.	-	
Panicum notatum Retz.	-	

+ = symptom appeared upon inoculation, - = symptom did not appear upon inoculation.

A wide range of plants found in rice-growing areas are served as weed hosts. Most reports appeared limited to host range to the Poaceae(Gramineae) or Cyperaceae within the Monocotyledonae. Most of the artifically inoculated plants harbour the pathogen in their roots. However by what means or how rapid the transmission from plant to plant may occur in the field under natural conditions is unclear. Also the extent of rice crop with pathogen free seed will get infection is unknown. Since the weeds may harbour the pathogen as well as the left over host plant debries in the soil, there is every possibility of serving as potential inoculum source.

Disease Cycle

The ShR pathogen, *S. oryzae* is well known as seed-borne as well as soil-borne pathogen. The estimation of disease transmissibility by seed was high in cvs TNI (75.3%) and RGS-20(83.7%) and the seedlings with severe infection die (IRRI, 1981).There are several causal organisms / agents which include rice insect pests namely rice borer, mite and mealy bugs are associated with ShR infection. Besides, a few fungal pathogens are also associated as co-pathogens with the ShR pathogens further, existence of rice tungro virus (RTV) on the host increases the ShR disease severity is also well known.

Chien and Chien (1964) reported the viability of the fungus was reduced considerably after 10 months at 15-28°C and completely lost after 12 months incubation. Raju (1980) and Singh and Raju (1981) further studied the survival of *S. oryzae* propagules in the infected tissue stored in soil under natural conditions. Spore concentration, temperature, soil pH, age of the culture and incubation period usually influence the survival of the pathogen as well as pathogenicity.

Singh and Raju (1981) reported that the causal organism survived up to 4 months in seeds and 7 months in leaf sheaths stored at room temperature. While it survived up to 10 months in sheaths incubated under field conditions. The fungus probably survives in soil on the stubbles of the infected plants left in the field and this may serve as initial source of inoculum to the next crop. However, there is still a lacuna in our knowledge of the disease cycle of the pathogen. In this context, recent reports of the posible collateral/ alternate hosts are interesting. The posibility of these weeds harbouring the fungus during the off-season and spreading the inocula to the later crop cannot be ruled out (Rahman et al. 1982).

A close scrutinity will have to be made to monitor the resting inocula of *S. oryzae* so that more effective control measures can be brought in. The events in the ShR disease cycle is given in Fig. 14 (Manibhushanrao, unpublished).

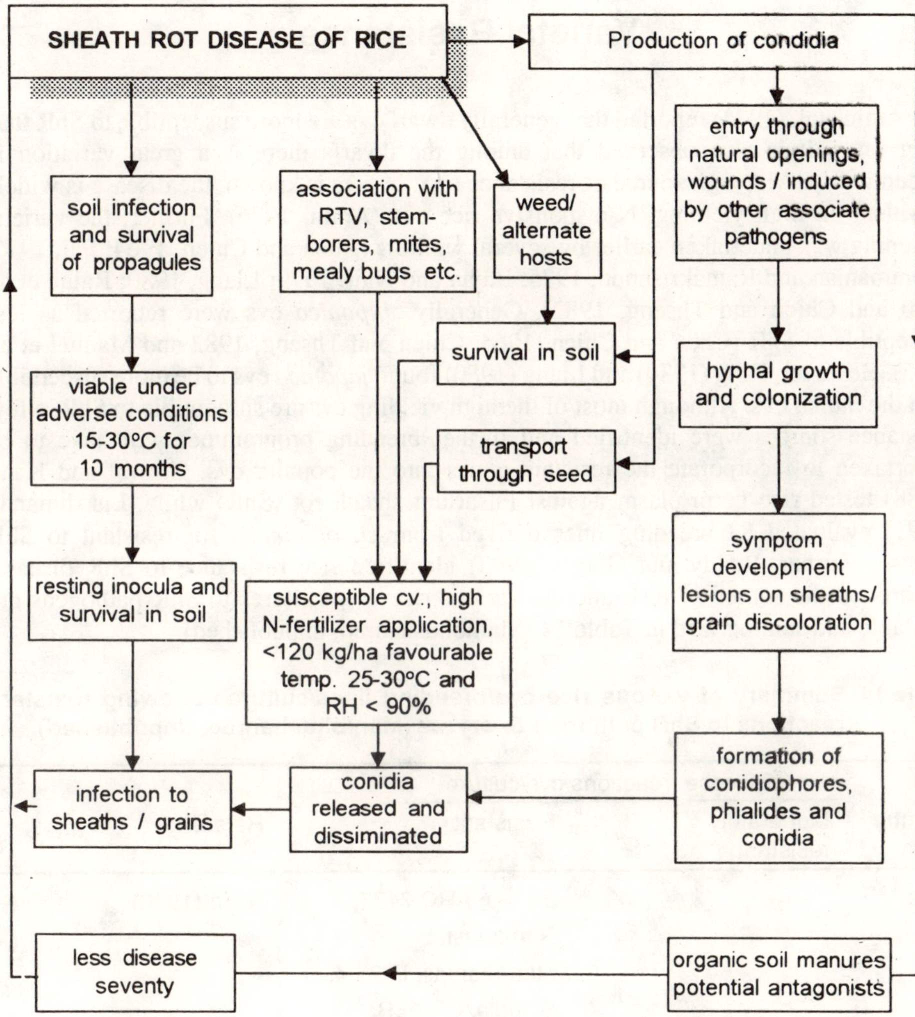

Fig. 14 : The key events of the disease cycle of *Sarocladium* sp. causing rice ShR disease (Manibhushanrao, unpublished).

Varietal Resistance

Amin etal. (1974) reported that generally dwarf cvs are more susceptible to ShR than taller ones. It is also observed that among the dwarfs there is a great variation in susceptibility. Although source of resistance to *S. oryzae* is known, the disease is widely prevalent on high yielding, N-responsive rice cvs (Amin, 1976). Further, the varietal screening was undertaken earlier by several workers (Chen and Chien, 1964; Pai, 1975; Subramanian and Ramakrishnan, 1975; Rajan and Nair, 1979; Liang, 1980; Raina et al. 1980 and Chien and Thseng, 1982). Generally, *Japonica* cvs were reported as less susceptible to ShR (Chen and Chien, 1964, Chien and Thseng, 1982 and Manuel et al. 1985). However, Fang (1980) and Liang (1980) found *japonica* cvs to be more susceptible than the *indica* cvs. Although most of the high yielding cvs are susceptible to ShR, a few resistance sources were identified and further breeding programmes will have to be undertaken to incorporate the resistant genes into the popular cvs. Grewal and Kang (1988) tested rice germ plasm against Fusarium sheath rot (ShR) while, Lakshmanan (1991) evaluated 87 breeding lines derived from *O. officinalis* for resistant to ShR disease. Further, Reddy and Ghosh (1993) identified rice resistance to ShR disease. Reports on the source of resistance to various rice cvs/ cultures to ShR pathogens are available and summarized in Table 14 (Manibhushanrao, unpublished).

Table 14: Summary of various rice cvs/breeding lines/cultures showing resistant reactions to ShR pathogen *S. oryzae* (Manibhushanrao, unpublished).

Country	Disease reactions cvs/culture		Reference
	Moderately resistant	Resistant	
India		Te tep, ARC 7177, Ramtulasi, Manoharsali, D-25-4, Sigadis, SR-26B, Intan, Zenith, Dissihatif, Raminad Str-3 and Tadukan	Amin (1976)
-do-		IR 24, IR 26, CR44-120-1, RP 825-71-4-1, RP 884-81-1 and RP 874-112-1-6	Naik (1976)

Country	Disease reactions cvs./culture Moderately resistant	Resistant	Reference
	AS 6979, 5821, 2992, 1287,3226, 6403,2692	—	Kannaiyan et al.(1978)
-do-	M23-17-1-1 M23-64-1-1	—	Balakrishnan et al.(1981)
	—	Mahasuri, Gajgor, Kala Namark, Usha Vishnubhog and Gajgor	Mukherjee and Singh (1980) Mukherjee et al. (1980)
-do-	Bas 370, Boro II, Chin Surah, TKM 9, Homothong and Sigadis	—	Raju and Singh (1981a)
		Pankaj	Misra and Mathur(1983)
-do-		Asha, Usha, Safari 17, Dubraj and Madhuri	Upadhyay and Diwakar(1984)
	TCA 190-1, RAU 83-8-4, BR-8, BR-34, Janaki, IET 6286	TCA 72, TCA 48, TCA 227	Singh et al. (1986)
-do-	—	Manoharasali RB 2069-2-6-1 RP 2071-18-1-1 CR 331-1-1-3 IR 92 02-21-1 RP 1057-35-1-1 VR S 286-4-1 ARC 14529 IET 8584	Lewin and Vidhyasekaran (1987b)
-do-	—	DP1 1091/5, white Ponni TAAU 80030, TAAU 80042TM 4300, TM8381, AU 42/1 P837, ACM1, AD85002 TM 10232, TKM 1, Ponni	Soroja et al. (1987)

Country	Disease reactions cvs./culture		Reference
	Moderately resistant	Resistant	
-do-	—	BPMS 69, BPMS 40A KalaNamak, KSR white	Singh(1988)
-do-	—	ADT 36, ADT 38, IR 50 ASD 7, ASD 8, ASD 17	Lakshmanan et al. (1991a)
		IR 54742-6-34-17-11-7* IR 54742-18-3-8-22-1* IR 54742-22-14-3-7-2* IR 54742-33-9-14-26-3* IR 54742-33-37-16-10-1 IR 54742-2-21-12-17-1 IR 54751-1-19-13-17-1* IR 54742-22-14-24-22-3* IR 54742-31-21-20-10-3 IR 54742-2-21-12-17-5	Lakshmanan et al. (1991b)
Taiwan	—	Taichung Sen Yu 229	Lin and Tsai (1977)

* Besides single grain inoculation, mealy bugs soaked in *S. oryzae* spore suspension was placed at the unemerged panicles.

** against *C. lunata* interaction on rice cvs.

Method of Inoculation

Inoculation of plants with spore-cum mycelial suspension either by spraying or injecting through the hypodermic needle of a syringe though a widely followed method, did not work out well with the rice-*S. oryzae* system. However, by inserting grain culture (2 wk old) either on the ligule, leaf sheath and stem or inside the flag leaf sheath ie., just above the floret on 62 d old plants gave 46, 97 and 93% ShR disease respectively (IRRI, 1977). This grain culture technique can be used on a large scale screening of the germplasm under field conditions. This study further indicated that ShR pathogen is not a weak pathogen and can insite disease in the absence of plant injury.

Several workers (Estrada et al. 1979; Raina and Singh, 1980 and Raju and Singh, 1981) studied the efficacy of different methods of inoculation and reported the superiority of insertion of rice/pearl millet grain culture into the flag leaf sheath over other methods produced typical symptoms. Estrada and Crill (1984) successfully induced ShR lesions with *S. oryzae* spore suspension prepared in 25% beef extract-peptone solution, when applied on plants treated with silicon carbide powder (60 girt carborundum) before inoculation. The ShR disease incidence was more severe as wounding greatly increased the disease development. Although, Kang and Rattan (1983) reported the development of disease symptoms with injection or insertion of mycelial cum spore suspension from PDA culture, insertion of grain culture inoculum seems to be the most feasible method as it produced ShR diseased lesions which would facilitate better scoring.

Method of Scoring

The reaction types (ShR lesions) of the infection on rice sheaths depends on the cv., virulence of the pathogens and other environmental conditions. The lesion types are used as the main criteria for evaluating varietal reaction. The size, shape and colour of the lesions on each leaf sheath are the general parameters considered.

The Standard Evaluation System (SES) (0-9) scale for rice suggested by International Rice Testing Programme (IRTP) (IRRI, 1988) is inadequate as the scoring is based on this system would only depict the disease prevalence and not the severity of a specific category. It also does not adequately identify resistant donors as the scale is based on the percent of ShR affected tillers. Earlier, Satyanarayana and Reddy (1979) designed a 1-8 scale portraying the coverage of lesions on the sheath and the damage to the size of panicle. The scale (Fig. 15) has adequate coverage of lesions on sheaths and their effect on the emerging panicle. This should appropriately identify resistant lines against ShR in the intensive rice growing areas especially in *rabi* and late *kharif* season crops. Further, Raychoudhury and Purkayastha (1980a) developed a scale based on the size of necrotic spots on the leaf sheath and Raju and Singh (1979) provided a (1-9) scale depending on the percentage of affected plants.

A standard scale for assesing ShR resistance in rice is quite essential to evolve and score uniform reactions to ShR pathogen. This will help in arranging the field data at large scale especially the IRTP entries from different countries uniformly even if there was a change in locality, environment and virulence of the pathogen. Spacing the seedlings should be uniform as it has positive correlation with N application and ShR incidence (Srinivasan, 1980b). Comparison of the the assessment methods is also essential as the area under the diseased progress curve (AUDPC) will be of use to compare the efficacy of the methods.

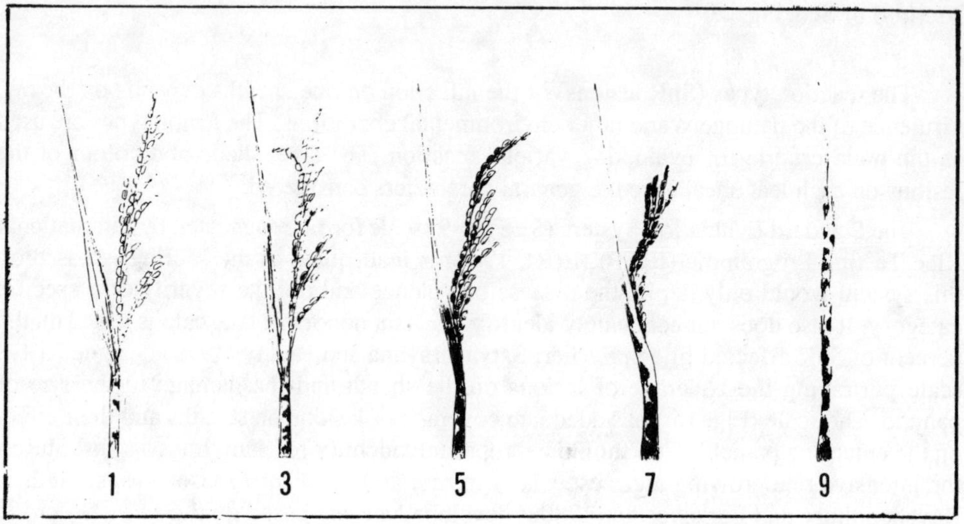

1. Resistant (R) = small brown lesions on the leaf sheath; normal panicle, emergence and grain setting normal.
3. Moderately resistant (MR) = lesions tend to enlarge, coalesce and cover about half of the leaf sheath; glumes are discoloured and grain filling normal.
5. Tolerant (T) = lesions tend to cover the entire sheath; glumes discoloured; incomplete exertion of the panicle with stray sterile spikelets.
7. Susceptible (S) = panicle struggles to emerge with 25-50% of spikelets sterile; severe glume discoloration and entire sheath rots.
9. Highly susceptible (HS) = no panicle emergence from the sheath; sheath rots while enclosing the panicle.

Fig. 15 : A simplified scoring system for rice ShR disease (Satyanarayana and Reddy, 1979).

Host -Pathogen Interaction

A sequential and indepth study would unravel the nature of the ShR pathogens as to how they affect the rice hosts and post-infectional biochemical changes during the host-pathogen interaction. There have been conflicting reports on the influence of host nutrition in relation to ShR disease development. Recent investigations showed a marked increase in the host susceptibility to ShR with increasing doses of N-nutrition. However, N- application as split dose in combination with potassium resulted in reducing infection.

It is well known that the ShR disease of rice is expressed by the development of characteristic irregular dark brown lesions on the sheaths (Plate 1ab: Chapter on Disease Development). This leads to the deranged host metabolism followed by reduction in biomass as well as yield of the rice plants. Hence, it is imperative to understand and assess, the physiological and biochemical factors responsible for these manifestations of the diseased host.

The development and spread of the ShR disease are mainly influenced by the virulence of the pathogen, environmental factors and susceptibility of the host, which in turn depend on nutritional status, genetic makeup and age of the rice plants. The *in vivo* cell wall degrading enyzmes have a major role in the possible mechanism(s) of entry and colonization of the pathogen into the host tissue.

Little information is available on the *in vitro* production of cell wall related enzymes of the ShR pathogens and most of the studies are not related to pathophysiology. Further, Joe and Manibhushanrao (1996b) analysed the changes in biochemical parameters governing resistance such as phenolics, phenol biosynthetic and oxidative enzymes in the diseased rice tissue under unfertilized and N-fertilized conditions. How far these attributes play a role to the host defence mechanism(s) and act against the invading ShR pathogens was emphasized.

N-fertilization

Akanda et al. (1984) reported that ShR has become a constrain to N- response on modern rice cv. BR 11 as well as for local improved cv. Nizersail from Dhaka, Bangladesh. A positive correlation between N-fertilization (0,30,60,90,20 Kg/ha levels) and ShR severity was established (Fig. 16). Subsequently, Jayasekhar and Prasad (1988) applied 5 levels of K and two levels of N in the 15 m x 15 m plots and the disease incidence (DI) was recorded 7 d before harvesting. The DI increased with increased levels of N and K with a constant level of P. Application of K significantly lowered the ShR incidence (Kanagasabai, 1980, Muralidhar, 1982). Again, NH_4Cl application lowered the ShR incidence significantly than with urea and $(NH4)_2$ SO_4 (Alagarsamy and Bhaskaran, 1986). Further, Joe (1994) studied the effect of N-fertilization and ShR infection on rice cvs CO 43 and White Ponni highly susceptible and resistant to ShR disease respectively.

The degree of susceptibility in cv. CO 43 increased due to high level of N-fertilization when infected with *Sarocladium* spp. showing high disease incidence (Fig. 17).

The level of reducing and non reducing sugars increased due to enhanced N applications in two rice cvs Kanagi and Bhavani which are susceptible and resistant to ShR respectively. The increase was higher in the susceptible cv. and tend to increase susceptibility to the invading pathogen (Mohan and Subramanian, 1977d).

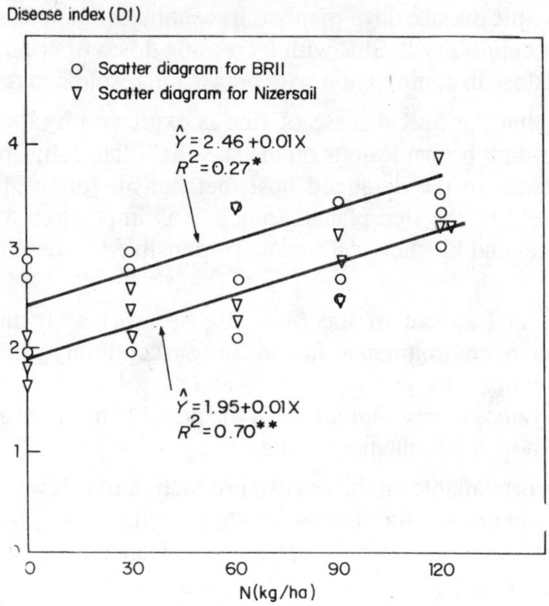

Fig. 16 : Relationship between N-fertilization and ShR severity (Akanda et al. 1984).

Joe (1994) reported that N-fertilization and ShR infection severely reduced chlorophyll(chl)- content in rice tissue . However, Chl-a is more sensitive than chl-b and was higer in N-fertilized uninoculated plants than unfertilized ones. Again, an increase in DI reduced the chl-b content to a greater extent in both the cvs tested and it was much lower than the control.

Application of N-fertilization and ShR infection increased the total sugar content in both the cvs but it was more in suscpetible cv. than in resistant one. Under fertilized condition, the total sugar content was more with the isolate SO 1 than SA 1. The reducing content was more in susceptible cv, than in resistant one. No corrleation was obtained between changes in reducing sugar content and days of inoculation. In general, N-fertilization had 2-fold increase in reducing sugar compared to unfertilized ones.

The starch content was also reduced on 22-DAI and sudden raise on 29-DAI after ShR infection (SA 1) particularly with N-fertilized susceptible plants. Such reduction could be attributed to the amylase activity of ShR pathogens. The SA 1 infected resistant cv. under N-fertilization showed both decreasing as well as increasing trends and this can be attributed to the pathogen at different stages of growth during infection.

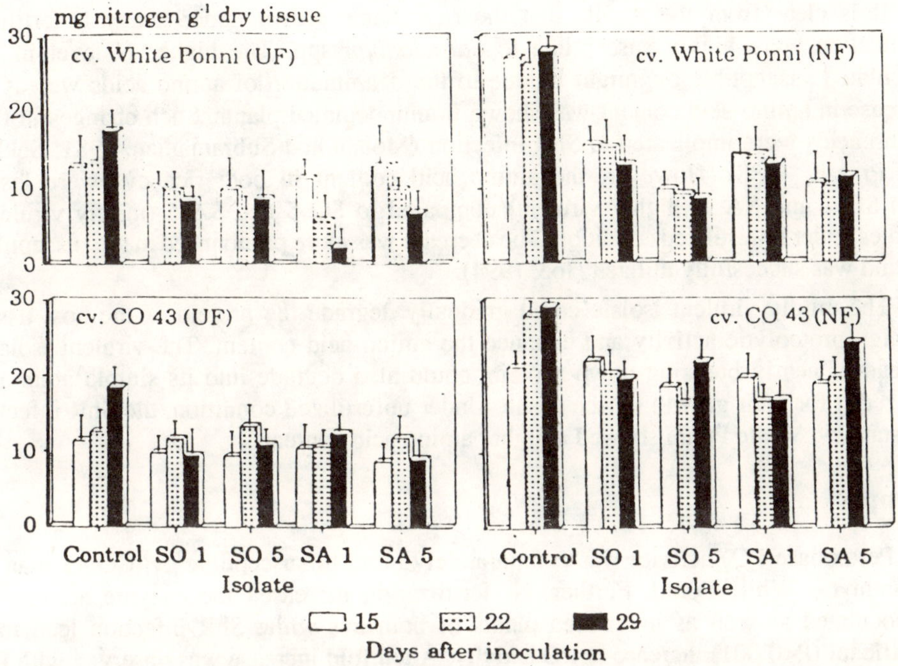

Fig. 17 : Changes in total nitrogen content in rice cvs influenced by N-fertilization and ShR infection, UF : Unfertilized NF : N-fertilized (Each value is represented the mean (±SE) of triplicates) (Joe, 1994).

Phenol Metabolism

Recently Joe and Manibhushanrao (1996b) studied the relationship between phenol metabolism and ShR disease resistance in rice. Total phenol levels increased in uninoculated resistant susceptible rice cvs. The SA 5 inoculated resistant cv. under unfertilized condition showed significant increase on 15-DAI when compared to SO 1 and SA 1. No definite trend was noticed in the total phenol with respect to disease progress. The uninoculated plants had more PAL activity (151-181 µg protein h^{-1})* than inoculated ones grown in N-fertilized plots. ShR infection drastically changed the PAL activity in all cases except unfertilized resistant cv. infected with isolate SO 5 (151-197 µg/ml) and SA 5 (277-295 µg/ml). The enyzme activity was more pronounced towards the initial stages of infection and gradually decreased thereafter with significant variation ($P < 0.001$).

In the case of TAL activity, it was less in inoculated cvs compared to uninoculated ones. There was a significant reduction ($P < 0.001$) due to N-fertilization in both the cvs. This was more pronounced in susceptible cv. TAL activity 70-90% was seen in both cvs. infected by SO 1 isolate under N-fertilized condition compared to respective control. The mildly virulent isolates SO 5 and SA 5 had a high enyzme activity. The reduction in TAL activity with increase in age was noticed in the infected plants (Joe and Manibhushanrao, 1996b).

It is clear from the results that the rice plants grown under heavy N-fertilizer application were highly susceptible to *Sarocladium* spp. The higher N level in the inoculated susceptible cv. might be due to the deamination of amino acids wereas an increase in amino acid content was shown in uninoculated plants. Such changes in free amino acids were implicated to ShR infection (Mohan and Subramanian, 1981; Sekhar and Prasad, 1988). However, the amino acid content in both rice cvs were lesser with SO 1 and SA 1 (highly virulent) compared to SO 5 and SA 5 (mildly virulent) isolates under N-fertilized condition. The decrease was more pronounced in the susceptible cv. and was succesfully utilized (Joe, 1994).

The mildly virulent isolate could gradually degrade the protein of the host tissue through proteolytic activity and enhance the amino acid content. The virulent isolate, instead of nearly breaking down protein, could also degrade into its simple units and utilized it for their growth in rice tissue. Under unfertilized condition, the ShR infected resistant cv. White Ponni showed a higher amino acid content.

PO/PPO

Peroxidase(PO) activity was comparatively more in susceptible cv. CO 43 than in resistant cv. White Ponni. Further, N- fertilization increased the enzyme activity in uninoculated as well as inoculated plants of both cvs while ShR infection lead to a significant ($P<0.001$) increase in PO activity. A ten fold increase was observed with the SO 1 infected resistant cv. and 2-3 fold increase in susceptible cv. The unfertilized resistant cv. infected with isolates SO 5 and SA 5 showed the lowest PO activity.

Polyphenol oxidase (PPO) was high in resistant cv. than in susceptible one. In general, PO and PPO levels were higher in resistant infected with isolates of *Sarocladium* spp. than the healthy plants. Maximum enzyme activity was detected in unfertilized resistant cv. infected by isolates SO 5 and SA 1. The increase PO activity may not be implicated in resistance as it was also observed in susceptible host tissue. The higher PPO activity could be used as a biochemical marker to predict resistance to disease in unfertilized resistant cv. infected with mildly virulent isolate SO 5 (Joe and Manibhushanrao, 1996b).

Plant Hormones

Raychoudhary and Purkayastha (1980a, b) reported the effect of plant hormones on the susceptibility of dwarf rice cvs of rice to ShR disease. Foliar application of gibberellic acid (GA_3) (50 and 100µg/ml) on 63-d old rice plants significantly reduced the ShR disease and altered the growth patterns of rice plants. Higher concentration of GA_3 increased amino acid level of two rice cvs Jaya (75%) and TN 1 (93%) initially but after 7 d of inoculation the concentration decreased considerably (Jaya, 30% and TN 1 32%).

The elicitation of momilactone by GA_3 treated healthy coleoptiles and leaf sheaths of rice was reported by Goshal and Purkayastha (1984). The susceptibility of the rice plants to ShR decreased when treated with GA_3 at 10 and 100 ppm. The GA_3 act as a precursor of gibberellin mediated enzyme (associated momilactone biosynthesis) production account for the elicitation of momilactone synthesis in GA_3 treated and non-inoculated rice tissue (Table 15). However, the mechanism involved in the induction of momilactone synthesis by GA_3 requires further study.

Table 15: Effect of GA_3 on momilactone A level in rice cv. Jaya (Ghosal and Purkayastha, 1984).

Treatment	Plant part	Concentration of momilactone A (μg/g fresh wt. rice tissue)	
		Healthy	Infected
Untreated	Coleoptile	0	5.6
Treated (10 ppm GA_3)	Coleoptile	13.2	16.7
Untreated	Leaf sheath	0	8.6
Treated (10 ppm GA_3)	Leaf sheath	14.5	19.9

Interaction With Other Stem Attacking Rice Pathogens

Gnanamanickam and Mew (1991) reported that some of the stem attacking fungal pathogens of rice viz., *Sclerotium oryzae* Catt. *Gaeumannomyces graminis* (Sacc.) von Arx and Oliver var. *graminis* Walker which cause stem rot (StR) and crown sheath rot (Cr.- ShR) respectively were found to be most sensitive to *S. oryzae*. On the otherhand, blast (Bl) pathogen (*P. oryzae* = *M. grisea*) and ShB pathogen (*R. solani*) were less sensitive. Further, ShR pathogen is not at all inhibitory to *Fusarium moniliforme* as well as *Xanthomonas campestris* pv. *oryzae*, the incitants of foot-rot and bacterial blight respectively.

Data from greenhouse studies showed that there was a reduction in the StR and cr-ShR by inoculation of these pathogens with *S. oryzae* (Table 16). This observartion was in support of the evidence obtained in *in vitro* studies. Further, in the presence of ShR pathogen the symptom development and disease severity of the StR and Cr-ShR pathogens are restricted. However, the ShR severity in the rice is not affected by the presence of ShB infection but increased the ShR severity in the presence of a pathogen sensitive to *S. oryzae*.

These results do suggest that ShR disease could emerge among the interactions of several stem attacking rice pathogens. It is further, indicated that symptoms of ShR and StR cannot exist on the same rice plant but ShR and ShB may occur together. Further, ShR pathogen reduced the severity of the diseases caused by the sensitive pathogen such as StR pathogen, *Sclerotium oryzae* by antibiosis and increases in disease severity in the presence of non-sensitive pathogen, *R. solani*.

These interactions helped ShR to become a dominant disease on rice but the actual mechanism(s) involved is not clearly understood. Although the outcome of this investigation is quite interesting, further extensive studies are needed to draw firm conclusions by making large number of systematic observations in natural occurrence in the field.

Table 16 : Effects of *Sarocladium oryzae* infection and its interactions with other fungal pathogens on crop growth and sheath rot severity in cv. IR 58 (Gnanamanickam and Mew, 1991).

Pathogen/pathogen combinations	Tiller number		Plant height (cm)		Biomass per plant (g)		Disease incidence mean lesion length (cm)					
							ShB		ShR		ShR%	
	I	II	I	II	I	II	I	II	I	II	I	II
I. No pathogen												
1. Uninoculated Check	12.8	15.4	84.0	81.0	73.4	74.8	0	0	0	0	21.0	30.0
II. Single pathogen												
2. *Sarocladium oryzae* (ShR)	17.4	19.6	66.5	69.0	56.1	61.0	0	0.8	0	0	34.2	58.0
3. *Sclerotium oryzae* (StR)	11.0	15.0	83.0	84.1	70.8	72.6	0	1.2	9.2	11.5	23.2	24.0
4. *Rhizoctonia solani* (ShB)	13.3	16.2	79.4	83.8	72.8	70.1	0.7	8.5	0.4	0	44.9	31.0
5. *Gaeumannomyces graminis var. graminis* (Cr-ShR)	15.4	16.0	78.7	80.9	72.9	71.0	0	0	0	0	21.8	24.0
III. Pathogen combinations												
6. ShR + StR	19.8	18.6	63.4	66.0	75.0	72.8	0	0	22	4.8	60.6	65.0
7. ShR + ShB	15.1	17.8	69.7	70.2	70.3	73.0	0.7	7.6	0	0	53.2	68.1
8. ShR + StR + ShB + Cr-ShR	14.2	19.1	69.8	67.8	66.5	73.0	0	3.5	7.5	7.0	49.0	53.0
9. ShR + ShB + Cr-ShR	17.4	16.0	79.6	78.4	73.1	74.5	0	1.0	7.2	13.8	19.6	26.0
LSD (0.05)	3.3	2.3	13.8	10.5	13.4	11.8	N.T	N.T	N.T	N.T	12.3	18.3

Greenhouse experiments (I and II), IRRI
Each value is an average of 4 replications
N.T. = not tested

Disease Resistance

There is enough evidence to show that disease resistance in plants is an active biochemical process. Phytoalexins and other post- infectionally accumulated anti-microbial substances play a crucial role in defence. The accumulation of oxidized phenolic compounds in rice which are toxic to major fungal pathogens is well known. Phytoalexin production especially momilactones in rice received increasing attention among rice researchers in recent years. Peroxidase (PO) and polyphenol oxidase (PPO) are well known in oxidizing phenols to quinones and the biosynthetic enzyme phenylalanine ammonia lyase (PAL) is evolved in the production of phenolic phytoalexin or prohibitins. Recently, the concept of induced resistance in rice has gained momentum. Earlier, Manibhushanrao and Zuber (1978) reviewed the role of anti-fungal and anti-microbial substances in disease resistance in cereals. The misconception that masked the approaches to plant control is poignantly summarised by Kuĉ (1987). Subsequently, Manibhushanrao et al. (1988) gave a comprehensive account on the changes in phenolic compounds and enzymes of phenolic biosynthesis and oxidization in host resistance against plant pathogens. It is essential to know the biochemical and molecular basis of resistance manifestation by plants to achieve efficient disease management. Further, the involvement of PO, PPO, PAL in rice-ShR pathogen system was studied (Raychoudhury and Purkayastha, 1980b; Rania and Singh, 1980 and Joe 1994).

Induced Resistance

Studies on the physiological alterations in the host in response to ShR pathogen, *A. oryzae* infection are very megare. Induction of resistance in rice plants against ShR disease was reported by Biswas and Purkayastha (1988). An increase in the total phenol content in rice leaf sheath up to 72 h irrespective of treatment of different systemic compounds viz., GA_3, aureofungin and bavistin or *A. oryzae* inoculated plants was observed. Maximum increase in phenol was noted in untreated, inoculated plants and minimum in GA_3 treated ones. A marginal decrease in phenol content has obtained in bavistin and aueofungin treated leaf sheaths after 72 h (Fig. 18). These results indicated that ShR pathogen caused significant changes in phenol content of rice plants. Further, Raychoudhury and Purkayastha (1980a,b) reported significant reduction in ShR incidence, increase of plant height and length of internodes in dwaf cvs , 'Jaya' and 'TN1' upon treatment with gibberellic acid but could not find any correlation between the observed resistance and the changes in amino acid and phenol pools of the host. The possible mechanism of the acquired resistance remains to be worked out.

Phytoalexins

Phytoalexins are the post infectionally formed antimicrobial substances play a crucial role in hosts' innate defence. The possible use of phytoalexins in plant disease

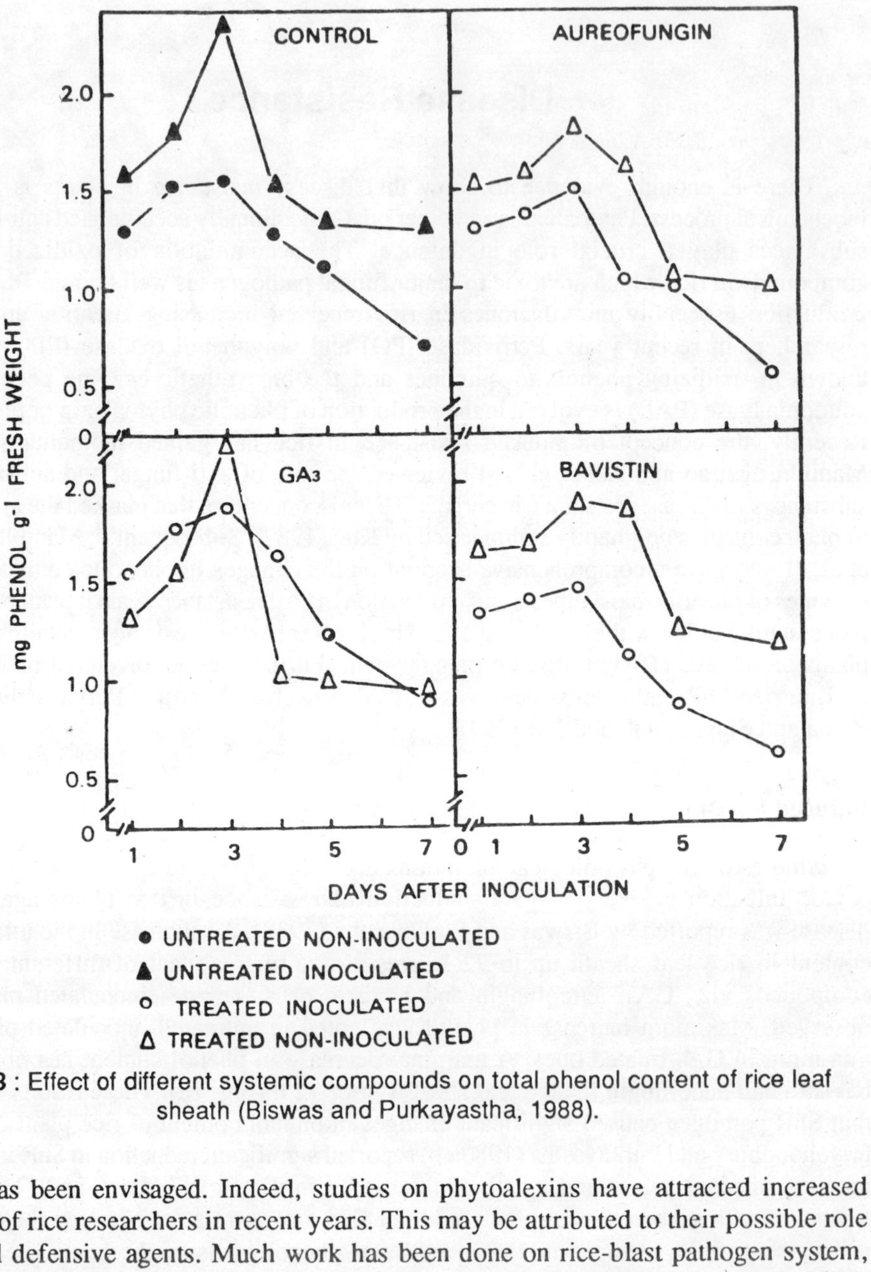

Fig. 18 : Effect of different systemic compounds on total phenol content of rice leaf sheath (Biswas and Purkayastha, 1988).

control has been envisaged. Indeed, studies on phytoalexins have attracted increased attention of rice researchers in recent years. This may be attributed to their possible role as natural defensive agents. Much work has been done on rice-blast pathogen system, however, a comprehensive account on the role of anti-fungal substances in rice-ShR pathogen *A. oryzae* is not available.

Purkayastha et al. (1983) reported the production of momilactone A - a diterpene lactone in the resistant tall rice cvs to ShR disease. Greater amounts of momilactone A were detected in *A. oryzae* infected dark grown coleoptiles different tall rice cvs while coleoptiles from semi-dwarf as well as ShR infected rice leaf sheaths showed less accumulation of the phytoalexin.

Purkayastha and Ghosal (1983) reported that the production of momilactone is associated with resistance of rice cvs to ShR disease. Tall rice cvs resistant to *A. oryzae* produced greater amounts of momilactone A (12.9-21.3 µg/g fresh wt.) while the semi-dwarf cvs produced relatively low amount (5.6-8.1 µg fresh wt.). The same trend was observed with momilactone A content of leaf sheaths of tall cv. Mashuri and semi-dwarf cv. Jaya (Table 17). It is pertinent to mention here that under field conditions tall rice cvs were found to be more resistant to *A. oryzae* than the susceptible semi-dwarf rice cvs (Amin et al. 1974).

Table 17 : Comparison of momilactone 'A' level in different rice cvs (Purkayastha et al., 1983).

Cultivar	Concentration of momilactone A in µg/g (fresh wt.)
Tall cvs (coleoptile)	
Badkalamkati	21.4
Mashuri	16.0
Rupsail	12.9
Mashuri(leaf sheath)	19.4
Semi-dwarf cvs(coleoptile)	
Jaya	5.6
CR-126-42-1	8.1
Ratna	5.9
Jaya(leaf sheath)	8.6

Momilactone A inhibited the germ tube growth of *A. oryzae* conidia when they have suspended in 'X' concentration (TLC fraction from 12g fresh weight of dark grown coleoptiles of rice cv. and rupsail disolved in 1ml of purified ethanol), while at higher concentration none of them showed germination (Purkayastha et al. 1983).

Further, the diffusates of tall and resistant cv. Mashuri and semi-dwarf and susceptible cv. Jaya were found to contain more fungitoxic substance in the former (74% germination of *A. oryzae* spores and 70% inhibition of germ tube lenght) than that of the later (12% inhibition of spore germination and 40% inhibition of germ tube length). However, no fungitoxic substance was detected either in leaf sheath exudates or in spore germination fluid (Purukayastha et al. 1983). These results indicated that resistant cvs contain higher amounts of momilactone A than the susceptible cvs irrespective of coleoptiles or leaf sheaths. However, it is not possible to conclude based on the results as more critical studies are needed with other momilactones-B,C,D produced by the rice cvs.

Ghosal and Purukayastha (1987) studied the responses of rice leaves to some abiotic elicitors. Among the six abiotic elicitors tested, GA_3, NaN_2 and pencillin induced the production of momilactone A in rice. The maximum production was obtained with GA_3 treated plants. These compounds also reduced the ShR disease incidence on rice (cv. Jaya) to some extent. One antigen was found missing following either treatment in Jaya. The application of GA_3 and NAN_3 significantly altered the common antigen relationship

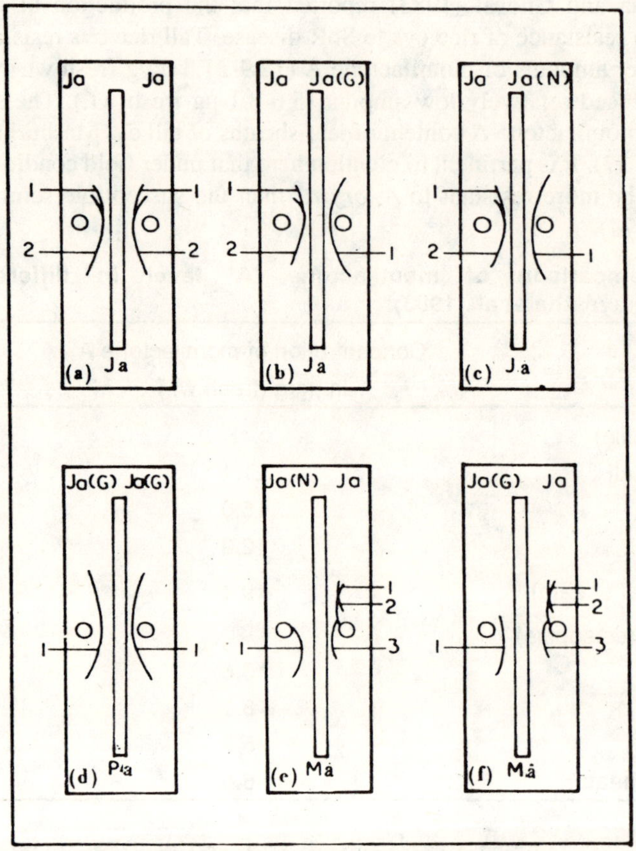

Fig. 19a-f : Immunograms showing immunoelectrophoretic patterns of rice antigens before and after chemical treatment and antigens of *S. oryzae*. Central rectangular troughs charged separately with antisera of cv. Jaya (JA) in (a), (b) and (c) and antisera of *S. oryzae* (PA) in (d) and Mashuri (MA) in (e) and (f). The surrounding wells charged with antigens of Jaya (Ja), Jaya treated with GA_3[Ja(G)], and Jaya treated with NaN_3[Ja(N)]. (Ghosal and Purkayastha, 1987).

between the host and the ShR pathogen, *S. oryzae* (Fig. 19 (a-f)). These results suggest that a close relationship might exist among phytoalexins, antigens, and disease resistance of rice plants.

Disease Control

The disease control is generally available through genetic (breeding for disease resistance) and use of chemicals and a combination of these two. Chemical contol is difficult because of the prohibitive cost of the pesticides as well as development of resurgence by the pathogens. Biopesticides as well as biocontrol agents using antagonists are effective and environmentally safe. Integrated disease management (IDM)-blending the triad viz. cultural, biological and the use of resistance source in the right manner could be adopted as there is great concern for the residual effect of pesticides to human, animal as well as plant health. Thus, the use of toxic chemicals can be avoided or minimized and the effective alternative is to be envisaged.

Breeding for Disease Resistance

In a few cases resistance to ShR disease was identified from breeding lines derived from *Oryza officinalis* (Lakshmanan and Velusamy ,1991). In other cases the level of resistance in the newly released cvs increased considerably. However, the spectrum of resistance against ShR was not broad enough to be stable. The dwarf cvs are considered as susceptible while the tall are always tolerant/resistant to ShR disease (Amin, 1974) and thus the tall ones are to be used in the breeding programme. Sreekumar et al. (1992) studied genetic evaluation of pre-released cultures/varieties of rice against ShR disease. Since the two pathogens *Sarocladium* spp. and other fungal co-pathogens cause typical ShR symptoms breeding for disease resistance is complicated and cumbersome. However, after successful screening of the germplasm a few cvs/lines were identified as resistant/ highly resistant to ShR pathogen (Table 12: Chapter on Varietal Resistance).

Chemical Control

If one looks back at the history of fungicides, copper fungicides enjoyed the pride of place in controlling most diseases in agricultural crops including rice. However, certain high yielding cvs of rice have found to be copper shy prohibiting the frequent use of these compounds. In other words, more copper fungicides were found to be slightly phytotoxic on certain rice cvs. Subsequently, the organic fungicides containing neither copper nor sulphur were introduced, which marked a great step forward in controlling many fungal diseases which include major rice diseases. Among them, the dithiocarbomate group such as dithane M-14, zineb, ziram, ferbem and maneb were the best known while maneb was very popular as had reached most prominant place among the fungicides in the middle of the century. Further, the development of mancozeb (dithane M-45) came next and is considered as the king of fungicides.

Later, with the introduction of systemic fungicides like benlate (benomyl), bavistin (carbendazim), busan (TEMBT), vitavax, kitazan, DDVP etc. Several years ago it was believed that the era of dithiocarbomate group will end. However, due to the limited activity and development of resistance in the fungal pathogens, the systemic fungicides failed miserably. Mancozeb retained its popularity as the new formulation (with 16% manganese and 2% zinc) has broad spectrum with fungicidal efficacy of more than 100 fungal pathogens has less phytotoxic and above all does not develop fungus resistance. It is also very effective against seed-borne ShR pathogens like *Sarocladium* spp.

Quite recently, attempts were made to evaluate the efficacy of certain fungicides and antifungal antibiotics in controlling rice ShR disease. Most systemic fungicides, the antibiotic aureofungin were effective in controlling the ShR disease (Mohan and Subramaniam, 1977a;Raghunathan and Vijayaraghavan,1976; Rajan and Nair, 1978; Chien and Huang, 1979; Raina and Singh, 1980 and Rajan, 1980). Foliar spray with different concentrations of benlate, MBC and vitavax showed that MBC was most effective fungicide (50mg/ml) in controlling ShR disease. Again, Kannaiyan (1979) reported reduction in disease intensity with two sprays, (10 d interval) of benlate, NF 48 and daconil at the flowering stage. Benthiocarb, a herbicide also reduced the severity of ShR disease (Vasavan et al. 1980). Five fungicides viz., difolaten (50%WP), Hoe 25986 (7%WP), benlate (50%WP), RH 2161 (2%EC) and deconil (75%WP) totally inhibited the spore germination of ShR pathogen (IRRI, 1978).

Further, polyoxin Z (Arunyanart at al., 1981), edifenphos and copper oxychloride (Govindarajan and Kannaiyan, 1982) were effective in reducing grain discolouration caused by *S. oryzae*. Rajan and Nair (1978) and Rajan (1980) studied the fungicidal control of ShR and glume discolouration in rice. Bavistin (0.05-0.1%), benlate (0.05-0.1%), brestanol (0.1%), dithane (0.4%), hinosan (0.1%) and difolaton (0.125%) were all effective in reducing the disease considerably (Raju and Singh, 1981b). Either a combination of tridemorph (100ml/ha) and bavistin (100g/ha) or tridemorph alone (200ml/ha) reduced ShR intensity significantly. Rajan (1981) attempted mechanical separation of seeds based on their specific gravity and reported lesser incidence of ShR with seeds separated in 25% salt solution. Separation of this method is simple, cheap and can be carried out rapidly by the farmers. Lakshmanan (1984) studied the efficacy of fungicides viz. carbendazim, tridemorph and mancozeb in a field trial for sheath rot control using cv. ADT 35. Tridemorph at 100 ml + carbendazim at 100 g/ha or the former at 200 ml/ha reduced the ShR intensity and increased the yield over the control significantly. Further, Alagarsamy and Bhaskaran (1986 a,b) reported that ShR disease incidence (DI) on cv. IR 20 was reduced significantly with the foliar application of $CaSO_4$, $ZnSO_4$, KCl, $CuSO_4$, $(NH_4)_2$ SO_4, NH_4 Cl and urea under greenhouse condition. Both $CaSO_4$ and $ZnSO_4$ sprays (0.2%) gave the highest grain yield while urea has no effect on DI on grain yield. On the other hand, NH_4Cl recorded the maximum yield (11.5 g/hill) with a DI of 35.8%. Alagarsamy et al. (1986a) found captafol (0.125%) had the lowest ShR severity (29.3%) and maximum yield (4.3 t/ha).

Reddy et al. (1983) studied the *in vitro* evaluation of different fungicides on ShR pathogen. Further, Lewin and Vidhyasekaran (1987a) reported that among the 12 fungicides, (carbendazim, edifenphos, mancozeb, thiophanate methyl, captafol, carboxin, glycophene, validamycin, IBP, iprodin, tridemorph and copper oxychloride) tested against ShR pathogen *S. oryzae* all of them inhibited mycelial growth *in vitro*. However, the fungicidal sprays at the panicle initiation of the high yielding cvs were ineffective aganist ShR infection.

Narasimhan et al. (1990) studied the effectiveness of different salts in controlling ShR incidence in greenhouse and in the field. Foliar application of maganese sulphate, calcium sulphate, ferrous sulphate and copper sulphate (10^{-2}M concentration), gave good protection for ShR disease in the field. Further, Narasimhan et al. (1994) reported that gypsum at 500kg/ha as a top-dressing single or in combination was comparable with carbendazim in reducing ShR and increasing grain yield significantly. These results do suggested that application of gypsum can reduce the ShR incidence as well as increase the grain yield.

Biological Control

Of late, biocontrol through the introduction of antagonistic microorganisms into rhizosphere has been recognized a successful method to combat soil-borne plant pathogens. So far, no effective attempts were made on the possible biocontrol of *S. oryzae* on lage field scale trials. The fact that ShR affects the crop at very late stages of the rice crop and the uncertainity of the disease cycle of *S. oryzae* makes biocontrol of ShR pathogen a difficult propostion. Nevertheless, an integrated approach to ShR control with chosen antagonists and soil amendments with organic manures would be a very exciting area for further research. Detailed information on the use of integrated control and the mechanism(s) involved to reduce the pathogen population in relation to biocontrol of rice sheath blight (ShB) disease is discussed elsewhere (Manibhushanrao, 1995).

Antagonists

Much emphasis was made in recent years on the biocontrol of plant pathogens through fungal antagonists (Manibhushanrao, 1995). Antagonist fluorescent psudomonads have also been used as biological control agents. The *Psudomonas fluorescens* and *P. putida* groups are collectively termed as plant growth-promoting rhizobacteria (PGPR). Sakthivel and Gnanamanickam (1986a) reported that *P. fluorescens* is antagonistic to *S. oryzae* and that seed bacterization with *P. fluorescens* (Pfc-strain) on rice plants (cv. TKM 9) under greenhouse and field studies could enhance plant height (12-27%) and substantially reduced the ShR infection thereby enhanced (3-16%) grain yield (Sakthivel and Gnanamanickam, 1986b). *P. fluorescence* decompose the organic phosphates as well as enhances the crop yield by reducing the inoculum density of soil borne plant pathogens. The specific mechanism(s) by which these bacteria promote the plant growth is not yet clear.

Phylloplane fungal isolates of *Bipolaris zeicola* obtained from IR 20 at tillering inhibited mycelial growth of *S. oryzae in vitro* and the undiluted culture filtrate completely inhibited the growth of the pathogen (Viswanathan and Narayanasamy, 1990). The efficacy of the potential antagonists of ShR should be tested on rice under field conditions as *in vitro* and *in vivo* studies may not be comparable.

Soil Amendments

Successful ShR disease control through soil amendment of industrial and agricultural wastes has been reported recently. Rajan (1980) studied the effect of soil amendment of various organic material and reported significant reduction in ShR intensity at 1 t/ha level with neem cake (*Azadirachta indica*), punna cake (*Callophyllum inophyllum*), marotti cake (*Hydrocarpus* sp.) and rubber seed cake (*Hevea brasiliensis*). Saw dust and mahua cake reduced the survival of the pathogen to a considerable extent (Kanagasabai, 1980). Organic soil amendments viz. neem cake, ground nut cake and gliricida leaf on the survival of ShR pathogen was attempted under greenhouse/field conditions and the treated soils produced less ShR infection and more healthier plants than the untreated ones (Manibhushanrao, unpublished). The beneficial effect may be either due to stimulation of soil saprophytes leading to the reduction in ShR pathogen propagules or better plant protection through increased soil fertility during the active decomposition period.

Biofertilizers

Application of bacterial biofertilizer *Azospirilluum lipoferum* along with N in the form of urea (0, 20 and 30 kg N/ha) reduced the ShR incidence. The treatment 20kg N + *A. lipoferum* recorded low ShR incidence (6.7%), increased grain yield (2.9 t/ha) which was 25.9% higher than 20 kg N/ha alone (Eswaramurthy et al. 1988).

Biopesticides

The biocidal properties of plant extracts are well known. The tropical flora which has to defend themselves against various types of microorganisms during the process of evolution are rich in antimicrobial substances. Biopesticides are less toxic to plants, systemic in nature, easily biodegradable but exhibit some phytotoxic effect at higher concentrations.

A number of compounds viz. azadiractin, mimosin and lawsone extracted from seeds of neem (*Azadiracta indica*), *Leucaena lucocephala* and leaves of *Lawsonia inermis* respectively have strong antimicrobial properties. They completely inhibit the growth of many phytopathogenic fungi including major rice pathogens viz. *P. oryzae*, *R. solani* and *Helminthosporium oryzae*. The commercial preparation of azadiractin is neemgold (contains 0.15% of it as active principle) is a tetranortriterpenoid compound while mimosine (β-(N-3-hydroxyl)- 4 pyridone - α - amino propionic acid) is a non-protein amino acid and lawsone (2-hydroxy-1, 4-napthoquinone) can be effectively used against ShR disease control.

Narasimham et al. (1993) tested the efficacy of a few plant products againh ShR pathogens and the preliminary investigations gave encouraging results. Most of these compounds can be exploited on a commercial basis, as biocides impair the metabolic function of the pathogens. More critical evaluation of biopesticides is needed with the field grown rice plants.

Remedial Measures

In recent years much emphasis was laid on ShR control through chemical, biological and agrochemicals as well as biopesticides. However, the general control methods recommended are as follows:

Since the pathogen is seed borne seeds can be soaked in bavistin (0.1%) for 30 min before sowing or dry seeds were treated with either thiram (2 g/kg seed) or captan (4g/kg seed) or dithane-45 (2-3 g/kg seed) will prevent the ShR disease.

Mechanical separation of seeds based on their specific gravity in 25% NaCl solution showed lesser ShR incidence and has been considered as inexpensive and efficient method.

ShR severity can be reduced by a application of foliar sprays with bavistin (0.05-0.1%), brestanol (1%), dithane (0.4%), hinosan (0.1%), deconil (0.2%), difoltan (0.125%) and MBC (50 mg/ml).

Foliar application of chemicals such as $CaSO_4$ (2%), $ZnSO_4$ (1%), KCl (1%) and $CuSO_4$ (0.2%) at 30 and 55 d after transplantation decreased the ShR incidence and increased the grain yield significantly.

The rice fields should be free from grass weeds and decayed stubbles in order to prevent the survival of the pathogen in the absence of rice crop.

Soil amendments with oil seed cakes of neem, punna, marotti and rubber will reduce the disease intensity to a greater extent.

Application of antagonistic *Psudomonas fluorescens* is an important biocontrol agent as it suppressed the pathogen propagules in the soil while seed bacterization reduced the ShR infection.

Integrated control such as a modest level of the host-resistance combined with other effective control measures viz., application of organic amendments, antagonists, biopesticides and vesicular arbuscular mycorrhizal (VAM) fungi in various combinations under upland conditions are the best available methods for an effective control of ShR disease.

Pathogen free seeds and good cultural practices are considered the ecologically sound tactics to control ShR disease.

Future Perspectives

Preamble

The total occurence of fungal diseases on rice in Asain countries (temperate, tropical and sub-tropical regions) is recorded as 37, while diseases caused by bacteria and viruses including mycoplasma will be 10 and 16 respectively. Blast and sheath blight are considered the principal disease of rice as they cause severe damage to the crop and widely distributed in irrigated and rainfed lowland ecosystem especially in Asia. Initially, ShR was considered as a minor one but it has received importance in recent years in view of heavy yield losses. Of late, its occurrence is very much felt and considered as potential threat to rice cultivation. The intensive cultural practices on high yielding semi-dwarf, N-responsive cvs lead to the failure of the crop against ShR disease.

The number of publications appeared on rice ShR as well as the pathogens indicated from rice database, IRRI for the years 1971-95 shows that maximum number of publications appeared during 1986-1990 but declined considerably in 1991-1995 (Fig. 20). This trend is the same with both on ShR disease as well as on fungal pathogens. However ,the publications on *Sarocladium* spp. appeared from 1976 onwards only. The list does not contain publications appeared in journals which are not listed in the database.

Host-Pathogen Interaction at Molecular Level

The study of host-pathogen interaction concerns the struggle between rice plants and the ShR pathogen has been incessant throughout their coexistence. Such interactions are often highly complex if the pathogen has co-pathogens as well as other casual agents. In this specific case the outcome depend on upon the presence or efficacy of the host's defence system. The rice plant may become diseased when the ShR pathogen(s) manage to invade it either unhindered pre-existing defence system or without eliciting the hosts induced resistance response(s). Any options of the disease may reflect the ability of the invading pathogen to overcome the plant's defence system(s).

Induced Host resistance

As on today very few reports exist on the induced resistance against field protection. However, induced resistance to fungal pathogens on rice using elicitors by altering the host metabolism under greenhouse condition are available. Protection of rice against brown spot pathogen *H. oryzae* (Hait and Sinha, 1986) and ShB pathogen *R. solani* (Manibhushanrao et al. 1990) using inducer chemicals is well known. Ideally, the inducers chosen should activate the host defence in the wake of the pathogen attack and not otherwise. Such elicitations in rice- ShR pathogen system should be considered.

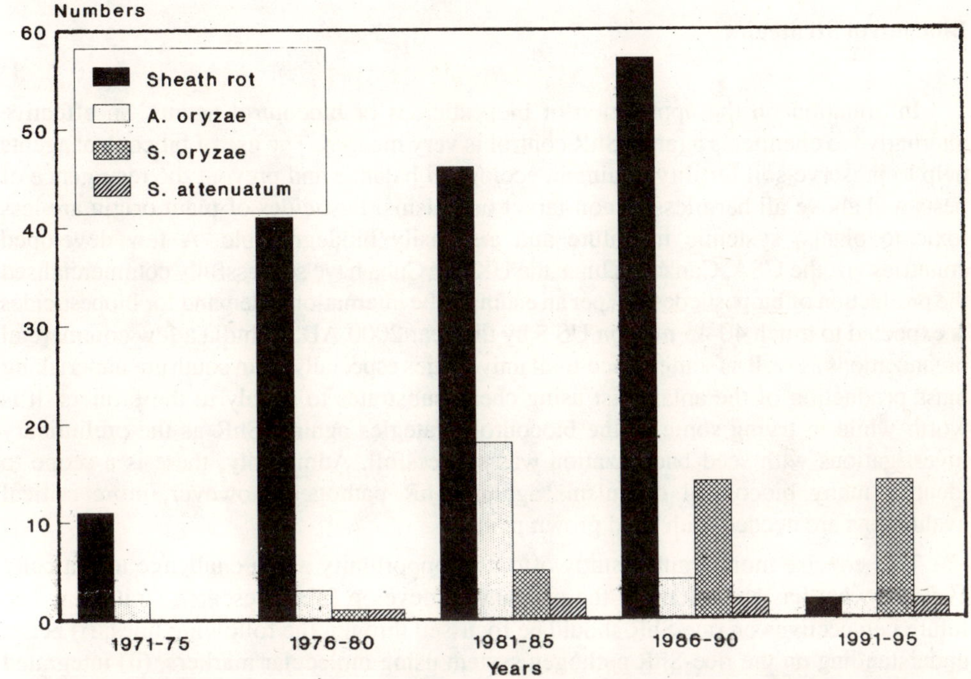

Fig. 20 : Number of publications appeared on rice sheath rot (ShR) disease and the incitants of ShR *S. oryzae* and *S. attenuatum* (Courtesy : IRRI Bibliographer, 1996).

Infact, virtually no work has been done on understanding the relationship of *Sarocladium* spp. and rice host at the molecular level. Restriction fragment length polymorphism (RFLP) analysis would reveal a better understanding of relationship among the ShR pathogen species. Identifying key determinant(s) of resistance in rice tissue against ShR pathogens along with rice tungro disease (association of ShR with tungro virus in the rice fields is well recognized) would be of great relevance in the present context. The pathogen related (PR) proteins /cell wall degrading enzymes viz., chitinase and ß, 1-3 glucanases are well known to degrade the fungal cell wall components. Hence, it is appropriate to analyze and identify such PR proteins in rice to manipulate natural resistance. The genetic and pathogenic diversity in rice ShR fungi can be studied further by isozyme polymorphism and DNA restriction analysis. The polymerase chain reaction (PCR) technique can be used to diagnose ShR fungus based on the amplification of the ribosomal DNA. This method is a convenient means of detecting sufficient amounts of DNA for molecular studies. Incidently, copathogens are involved in the ShR disease syndrome and PCR application may be very useful for screening and identifying potential pathogens.

Further, the split application of N-fertilizers in combination with the other organic amendments would probably alleviate the direct effect of N-accumulation and thereby reduce the susceptibility of rice plants. Over the years it has been a common practice for the breeders to introduce the resistant genes one after another, whereas the pathogens appeared to develop resistance and the resistance becomes ineffective after the development of a new race. Thus, newer strategies should be evolved for the control of ShR disease.

Biocontrol Strategies

Information on the application of biopesticides or biocontrol agents, an effective alternative to chemicals against ShR control is very meagre. The use of biocontrol agents help to preserve soil fertility, maintain ecological balance and prevent the resurgence of pests and above all harmless to non-target organisms. Pesticides of plant origin are less toxic to plants, systemic in nature and are easily biodegradable. A few developed countries viz. the USA, Canada, China ,the UK and Cuba have successfully commercialised the production of biopesticdes. As per an estimate the international demand for biopesticides is expected to touch 40-45 million US $ by the year 2000 AD. In India a few commercial organizations as well as some agricultural universities especially from south are undertaking mass production of the antagonist using cheap substrates to supply to the farmers. It is worth while in trying some of the bioconrol strategies against ShR as the preliminary investigations with seed bacterization was successfull. Admittedly, there is a scope to identify many biocontrol organisms against ShR pathogen, however, more critical evaluations are needed with field grown plants.

This concise monographic study offers an opportunity and a challenge to critically evaluate what has been done in the past and to develop future research strategies. The future perspectives on rice ShR should be focussed through the following lines; (i) better understanding on the rice-ShR pathogen system using molecular markers, (ii) integrated disease management and (iii) using induced resistance technology.

References

Agnihothrudu, V. 1973. *Acrocylindrium oryzae* Sawada-sheath rot on paddy. *Kavaka*, **1**: 68-71.

Ahamed, N.J., S. Kannaiyan and A. Venkatarao. 1975. First record of *Acrocylindrium*, sheath rot disease of rice in India. *Sci. Cult.* **41**: 544-545.

Akanda, S.I., A.K.M. Shahjahan, and S.A. Miah. 1984. Nitrogen fertilization and sheath rot (ShR) development in rice. *Int. Rice Res. Newsl.* **9**(6): 12-13.

Alagarsamy, G. and R. Bhaskaran. 1986 a. Effect of foliar spray of macro and micronutrients on sheath rot(ShR) incidence and rice grain yield. *Int. Rice Res. Newsl.* **11**(4): 27-28.

Alagarsamy, G. and R. Bhaskaran. 1986 b. Effect of form and level of N fertilizers on sheath rot(ShR) incidence and yield. *Int. Rice Res. Newsl.* **11**(6): 18.

Alagarsamy, G., D. Alice, and P. Palanisamy. 1986a. Chemical control of brown spot(BS) and sheath rot(ShR) in Tamil Nadu *Int. Rice Res. Newsl.* **11**(4): 18.

Alagarsamy, G., M. Vellusamy, S. Rajagopalan, and S. Palanisamy. 1986b. Effect of slow release N fertilizers on stem borer (SB) and sheath rot (ShR) incidence and on rice grain yield.*Int. Rice Res. Newsl.* **10**(4): 19.

Amin, K.S., B.D. Sharma and C.R. Das. 1974. Occurrence of sheath rot of rice in India caused by *Acrocylindrium. Pl. Dis. Reptr.* **58**: 358- 360.

Amin, K.S. 1976. Sources of resistance to *Acrocylindrium oryzae*, Sheath rot of rice. *Pl. Dis. Reptr.* **60**: 72-73.

Anima, P. 1975. Response of some rice varieties towards sheath rot (*Acrocylindrium oryzae*) infection. *Rice Path. Newsl.* **2**: 7.

Anonymous. 1962. Studies on plant diseases. *Crop Improve. Prog.* **2**:289- 295.

Anonymous. 1975. Testing varietal reactions and culture filtrate of the causal fungus. *Korean J. Pl. Prot.* **14**: 23-27.

Arison, B.H. and S. Omura. 1974. Revised structure of cerulenin. *J. Antibiot.* **27**: 28-30.

Arunyanart, P., A. Surin, S. Disthaporn. 1981. Seed discolouration disease and its chemical control. *Int. Rice Res. Newsl.* **6**(3): 14-15.

Attabhanyo, A. and M.C. Rush. 1973. Sheath rot of rice in Louisiana. In: *Proc. 2nd Int. Cong. Pl. Path.* Minneapolis, USA, pp.5-12.

Balakrishnan, B. and M.C. Nair. 1981. Weed host of *Acrocylindrium oryzae* Saw, sheath rot pathogen of rice. *Int. Rice Res. Newsl.* **6** (6):13.

Balakrishnan, B., K.M. Rajan, G. Mathai, N. Remabai. 1981. Disease reaction of advanced rice cultures. *Int. Rice Res. Newsl.* **6**(4):5-6.

Berdy, J. 1986. Further antibiotics with practical application In: *Biotechnology*. H. Pape and H.J. Rehn (eds). *Microbial Products* II**4**: 465-501.

Biswas, S. and R.P. Purkayastha. 1988. Induction of resistance in rice plants against sheath rot disease. *Ind. Phytopath.* **41**: 51-56.

Boa, E.R. and B.L. Brady. 1987. *Sarocladium oryzae* associated with blight of *Bambusa* species in Bangladesh. *Trans. Br. mycol. Soc.* **89**: 161-166.

Boa, E.R. and M.A. Rahman. 1987. Bamboo blight and the bamboos of Bangladesh. *Forest Path. Ser. Bull.* No.1 Forest Res. Inst., Chittagong.

Bonorden, H. 1851. Handbuch der allgemeinen Mycologie. Schweizerbart'sche Verlagshandlung, Stuttgart.

Brady, B.L.K. 1980a. *Sarocladium oryzae*. CMI descriptions of pathogenic fungi and bacteria. No.673. CAB Intl. Mycol. Inst., Egam, Surry, UK.

Brady, B.L.K. 1980b. *Sarocladium attenuatum.* CMI descriptions of pathogenic fungi and bacteria. No.674. CAB Intl. Mycol. Inst., Egam, Surry, UK.

Bridge, P.D., D.L. Hawksworth, D.F. Kavishe and P.A. Farnell. 1989. A revision of the species concept in *Sarocladium,* the causal agent of sheath-rot in rice and bamboo blight, based on morphometric analyses. *Pl. Pathol.* **38**:239-245.

Carrole, R.B., F.I. Lukezic and R.G. levine. 1972. Absence of common antigen relationship between *Corynebacterium insidiosum* and *Medicago sativa* as a factor in disease development. *Phytopathology,* **62**: 1351-1360.

Chakrabarti, D.K. and S. Biswas. 1978. Estimation of yield loss in rice affected by sheath rot. *Pl. Dis. Reptr.* **62**: 226-227.

Chakrabarthy, B.N. and R.P. Purkayastha. 1983. Serological relationship between *Macrophomina phaseolina* and soybean cultivars. *Physiol. Pl. Pathol.* **23**: 197-205.

Charudattan, R. and J.E. DeVay. 1972. Common antigen among varieties of *Gossypium hirsutum* isolates of *Fusarium* and *Verticillum* species. *Phytopathology,* **62**: 230-234.

Chen, C.C. and C.C. Chien. 1964. Some observations on the outbreak of rice sheath rot disease. *J. Taiwan agric. Res.* **13**:39-45.

Chen, M.J. 1957. Studies on sheath rot of rice plant. *J. Agric. For. Taiwan,* **6**: 84-102.

Chien, C.C. and C.L. Chu.1970. Studies on the effect of fertilizers to rice diseases. *J. Taiwan agric. Res.* **19**: 62-71.

Chin, K.M. 1974. Sheath rot disease of rice. *MARDI Res. Bull.* **2**: 9-12.

Chien, C.C., C.H. Huang. 1979. The relation between sheath rot and the sterility of rice plant. *J. agric. Res. China,* **28**: 7-16.

Chien, C.C. 1980. Sheath rot disease and its relation to the sterility of rice plants. *Pl. Prot. Bull. Taiwan,* **22**: 31-40.

Chien, C.C., F.M. Thseng. 1981. Physiological and biological studies of rice sheath rot pathogen. *J. agric. Res. China,* **30**: 1-4.

Chien, C.C., F.M. Thseng. 1982. Rice variety and age in relation to the susceptibility to sheath rot. *J. agric. Res. China,* **31**:155-161.

Chow, Y.S., S.S. Tzean, C.S. Chang, C.W. Wang. 1980. A morphological approach of the tarsonemid mite *Steneotarsonemus spinki* (Tarsonemidae) as a rice (*Oryza sativa var. spontonea*) plant pest. *Acta Arachn.,* Tokyo, **29**: 25-42.

Chung, H.S. 1975. Studies on sheath rot of rice caused by *Acrocylindrium oryzae* Sawada. Testing varietal reactions and cultural filtrate of the causal fungus. *Korean J. Pl. Protect.* **14**: 23-27.

Dachene, W.V., H. Lorch and W.O. Godtfredsen. 1968. Microbiological transformation of fucidana-type antibiotics-a correlation between fucidic acid and helvolic acid. *Tetrahedron letters* **47**: 4843- 4846.

Datnoff, L.E., M.L. Elliot, D.B. Jones. 1993. Black sheath rot caused by *Gaeumannomyces graminis* var. *graminis* on rice in Florida. *Pl. disease.* **77**: 210.

Datta, S.K. and R.P. Purkayastha. 1978. Germination behaviour of *Acrocylindrium oryzae* causing sheath rot disease of rice. *Indian Phytopath.,* **31**: 89-91.

Deka, A.K. and A.K. Phookan. 1992. Some common weed hosts of *Sarocladium oryzae* in Assam in India. *Int. Rice Res. Newsl.,* **17** (6): 25.

Detry, J.F. 1993. Seed dry-heat treatment against transmission of *Pseudomonas fuscovaginae,* causal agent of bacterial sheath brown rot of rice (BSR). *Int. Rice Res. Inst. Newsl.* **18** (2): 27-28.

Estrada, B.A. and J.P. Crill. 1980. Amendments to *Acrocylindrium oryzae* spore suspension to induce sheath rot infection of rice. *Int. Rice Res. Newsl.,* **5** (5): 8-9.

Estrada, B.A., L.M. Sanchez, J.P. Crill. 1979. Evaluation of screening methods for sheath rot resistance of rice.*Pl. Dis. Reptr.,* **63**: 908- 911.

Estrada, B.A., C.Q. Tores and J.M. Bonman. 1984. Effect of sheath rot on some rice yield components. *Int. Rice Res. Newsl.,* **9** (2): 14.

Eswaramurthy, S., M. Muthusamy, S. Mariappan, V. Kumar K. Jayasekar and S. Natarajan 1988. Management of rice brown spot, sheath rot and bacterial leaf blight with *Azospirillum lipoferum* application. National Seminar on Management of crop diseases wih plant products/biological agents, (Abstr.), p. 8.

Fang, H.C. 1980. Etiological studies on the sterility of rice plant and its control. *Pl. Prot. Bull.*, **22**: 83-90.

Gams, W. and D.L. Hawksworth. 1975. The identity of *Acrocylindrium oryzae* Sawada and a similar fungus causing sheath rot of rice. *Kavaka*, **3**:57-61.

Ghosal, A. and R.P. Purkayastha. 1981. A brief review of sheath rot disease of rice. *IRC Newsl.*, **30**: 34-36.

Ghosal, A. and R.P. Purkayastha: 1984. Elicitation of momilactone by gibberelline in rice. *Curr. Sci.*, **53**: 506.

Ghosal, A. and R.P. Purkayastha. 1987. Biochemical response of rice (*Oryza sativa* L.) leaves to abiotic elicitors of phytoalexin. *Ind. J. Expt. Biol.*, **25**: 395-399.

Ghufran, S.M., S.M. A. Asghar and A.P. Singh. 1980. Sheath rot spread in Bihar, India. *Int. Rice Res. Newsl.* **5** (5):11.

Gill, M.A. 1993. Sheath rot (ShR) disease of rice in Punjab, Pakistan, *Int. Rice Res. Newsl.*, **18** (2): 23.

Gill, M.A., S.I. Yasin, T. Khan, Q. Ashan Ullah and M.A. Butt. 1993. Sheath rot (ShR) disease of rice in Punjab, Pakistan. *Int. Rice Res. Newsl.* **18** (2): 23.

Gnanamanickam, S.S. and T.W. Mew. 1988. Interactions between *Scarocladium oryzae* and other fungal pathogens of rice. *Pl. and Soil*, .**138**: 213-219.

Gopinath, R. 1986. Studies on the epidemiology and ecology of sheath rot disease of rice caused by *Acrocylindrium oryzae* Sawada. Ph.D. Thesis, Osmania Univ., Hyderabad, India, p.210.

Govindarajan, K. and S. Kannaiyan. 1982. Fungicide control of rice grain infection. *Int. Rice Res. Newsl.*, **7**(1): 10.

Growth, D. 1992. Sheath rot. In: *Compendium of Rice Diseases*. (eds. R.K. Webster and P.S. Gunnel). APS Press, St. Paul, Min. p.25.

Grewal, S.K. and M.S. Kang. 1988. Screening rice germplasm against Fusarium sheath rot (ShR) disease. *Int. Rice Res. Newsl.*, **13** (3): 14.

Hait, G.N. and A.K. Sinha 1987. Biochemical changes associated with induction of resistance in rice seedlings to *Helminthosporium* infection by seed treatment with chemicals. *Z. Pfkrankh. Pflschutz* **94**: 360-368.

Harahap, Z., A.K.M.Shahjahan and M.C. Rush. 1974. Sheath rot of rice in Louisiana. *Proc. 15th Rice Tech. Working Group Meeting*, Univ. Ark. Fayetteville, Ark, p.45.

Hashioka, Y. and M. Makino. 1956. Relation of nitrogen nutrition of rice plants to the susceptibility to four foliage diseases. *Res. Bull. Fac. Agric. Gifu Univ.*, **6**: 58-66.

Hashioka, Y. 1963. Rice blast varietal resistance in relation to the local environments with notes on stem nematode and other diseases of rice in Thailand. *Final Rep. F.A.O.* p.66.

Hsieh, S.P.Y., W.L. Liang and S.Y. Chang. 1977. Etiological studies on the sterility of rice plants. I. Association of sheath rot fungus *Acrocylindrium oryzae* Sawada with sterile rice plant. *Pl. Protect. Bull. Taiwan*, **19**: 30-36.

Imolehin, E.D. 1989. Rice diseases in Southern Guinea Savannah Zone of Niger State Nigeria. *Int. Rice Res. Newsl.*, **14** (3): 37-38.

IRRI, 1978. A. Rep. 1979.

IRRI, 1977. A. Rep. 1978.

IRRI, 1981. A. Rep. 1982.

IRRI. 1988. Standard evalution system for rice (3rd ed.). P.O. Box, 933, Manila, Philippines, p.54.

Iwai, Y., J. Awaya, T. Kesado, H. Yamada, S. Omura and T. Hata. 1973. Selective production of cerulenin by *Cephalosporium caerulens* KF- 140. *J. Ferment. Technol.*, **51**: 575-581.

Jayasekhar, M. and N.N. Prasad. 1988. Effect of N and K on rice sheath rot (ShR) and crop yield. *Int. Rice Res. Newsl.* **13**(4): 38.

Jimenez, S. and P.S. Carbos. 1977. Rice sheath rot in Peru. *Int. Rice Res. Newsl.*, **2** (6): 12

Joe, Y. 1994. Studies on sheath rot disease of rice. Ph.D. thesis, Univ. of Madras, Madras, India. p. 127.

Joe, Y. and K. Manibhushanrao. 1995. Biochemical constituents of differentially virulent isolates of *Sarocladium* spp. causing sheath rot disease of rice. *Z. Pfschutz. Pflkrankh.* **102**: 291-198.

Joe, Y. and K. Manibhushanrao. 1996a. Morphological and biochemical comparison of *Sarocladium oryzae* and *S. attenuatum* causing sheath rot in rice. *J. Biosci.* (In press).

Joe, Y. and K. Manibhushanrao. 1996b. Relation between phenol metabolism and sheath rot resistance in rice. *Acta Phytopath Hung.*(In press).

Joe, Y. and K. Manibhushanrao. 1996c. Protein profile and serological relationship of *Sarocladium* spp. (In press).

Kalyanasundaram, R., C. Lakshminarasimhan and S. Venkataraman. 1975. Common antigens in host-parasite relationship. *Curr. Sci.*, **44**: 55- 56.

Kanagasabai, R. 1980. Studies on an integrated approach for the control of sheath rot disease of rice caused by *Acrocylondrium oryzae* Sawada. M.Sc. (Ag.) thesis, Annamalai Univ. Annamalainagar India, p. 82.

Kang, M.S. and G.S. Rattan. 1983. Sheath rot in the Punjab, India. *Int. Rice Res. Newsl.*, **8** (3): 7-8

Kannaiyan, S., V. Jayaraman and V.G. Palaniyandi. 1978. Reactions of certain rice cultivars to sheath rot disease. *Int. Rice Res. Newsl.*, **3** (1) :7

Kannaiyan, S. 1979. Chemical control of sheath rot disease of rice. *Int. Rice Res Newsl.*, **4** (3): 14-15.

Kawamura, E. 1940. Notes on the sheath rot of rice plants with special reference to its causal organism *Acrocylindrium oryzae* Sawada. *Ann. Phytopathol. Soc. Japan*, **10**: 55-59.

Kimura, H. and S. Mizyhima. 1973. An exopolygalacturonase from a strain of *Acrocylindrium. Agric. Biol. Chem.*, **37**: 2589-2593.

Koteswararao, D. 1977. Incidence of sheath rot on three rice varieties. *Int. Rice Res. Newsl.*, **2** (5): 16.

Kuč, J.1987. Plant immunization of cucurbits for disease control. In: *Innovative approaches to plant disease control* (ed.I. Chet) pp. 255-275. Wiley & Sons, NY. p. 372.

Lakshmanan, P. 1984. Effective control of sheath rot (ShR) disease. *Int. Rice Res. Newsl.*, **9** (5): 14.

Lakshmanan, P. 1991. A new sheath rot (ShR) disease of rice identified in Tamil Nadu. *Int. Rice Res. Newsl.*, **16** (5): 20.

Lakshmanan, P. 1992a. Effect of fungicides, insecticides, and their interaction on sheath rot(ShR) severity. *Int. Rice Res. Newsl.*, **17** (2): 14.

Lakshmanan, P. 1992b. *Brevennia rehi* Lindinger, vector for the sheath rot (ShR) pathogen. *Int. Rice Res. Newsl.*, **17** (2): 23.

Lakshamanan, P. and R. Velusamy. 1991. Resistance to sheath rot (ShR) of breeding lines derived from *Oryza officinalis*. *Int. Rice Res. Newsl.*, **16** (6): 8-9.

Lakshamanan, P., S.M. Kumar and R. Velusamy. 1992. Sheath rot (ShR) severity due to rice bug infection. *Int. Rice Res. Newsl.*, **17** (3): 23.

Lakshamanan, P., S.M. Kumar, R. Velusamy and K. Indira. 1991. Loss of rice grain yield and seedling vigor due to sheath rot (ShR) and mealy bug interaction. *Int. Rice Res. Newsl.*, **16** (6): 27.

Lewin, H.D. and P. Vidhyasekaran. 1987a. Controlling sheath rot ShR in rice. *Int. Rice Res. Newsl.*, **12**(1): 17.

Lewin, H.D. and P. Vidhyasekaran. 1987b. Sources of resistance to sheath rot (ShR). *Int. Rice Res. Newsl.*, **12**(1): 7.

Liang, W.J. 1980. Rearing method for tarsonemid mites (*Steneotarsonemus spinki*) free of rice sheath rot (*Acrocylindrium oryzae*). *Pl. Prot. Bull. Taiwan*, **22**: 23-30.

Lin, T.F. and T.T.Tsai. 1977. *Indica* rice varietal improvement for major diseases and insects resistance. *J. agric. Ass. China, News Ser.*, **102**: 86-95.

Lo, K. and C.C. Ho. 1979. Studies on the cause of the empty head of rice in Taiwan. *J. agric. Res. China*, **28**: 193-197.

Maiti, D., V.D. Shukla and M. Variar., 1992. A linear predicting model for assessing yield loss in upland rice due to sheath rot. *Oryza* **29** (2): 173-174.

Malathy, A. 1984. Physiological studies on *Acrocylindrium oryzae* the causal organism of sheath rot of rice. M.Phil. Dissertation Univ. Madras, Madras, India. p.64.

Manibhushanrao, K. 1994. *Rice Blast Disease*. Daya Publishing House, Delhi. p.180.

Manibhushanrao, K. 1995. *Sheath Blight Disease of Rice*. Daya Publishing House, Delhi. p. 101.

Manibhushanrao, K. and S. Sreenivasaprasad. 1985. Inhibition of polygalacturonase activity of rice pathogens by the oxidized products of chlorogenic acid. *Curr. Sci.* **54**: 146-148.

Manibhushanrao, K., R. Kalyanasundaram and S. Suryanarayanan. 1971. Antigenic comparison of cultures of *Pyricularia oryzae* Cav. *Phytopath. Z.*, **70**: 121-126.

Manibhushanrao, K., Y. Joe and P. Madathiammal, 1990. Elicitation of resistance in rice to sheath blight. *Int. J.Trop. Plant Dis.* **8** : 193-197.

Manibhushanrao, K., S. Sreenivasaprasad and R.S. Chitralekha. 1986. Sheath rot disease of rice. *Z. Plfschutz. Pflkrankh.*, **93**: 319-329.

Manibhushanrao, K. and M. Zuber and N. Matsuyama. 1988. Phennol metabolism in plant disease resistance. *Acta Phytopath. et. Entomol. Hung.* **23**: 103- 114.

Manibhushanrao, K. and M. Zuber. 1978. Disease resistance in cereals. *Acta Phytopathol. Acad Scient. Hungaricae* **13**: 113-135.

Manuel, W.W., K. Ganesan, C.K. Rajagopalan and V. Mariappan, 1985. New source of resistance to major rice diseases. *Int. Rice Res. Newsl.* **10** (1) : 7.

Mathur, S.C. 1981. Observations on diseases of dryland in Brazil. *Int. Rice Res. Newsl.*, **6** (6): 11.

Mew, T.W. and A.M. Rosales. 1984. Relationship of soil microorganisms to rice sheath blight development in irrigated and dryland rice cultivars. In: *Soil-borne crop diseases in Asia*, FFTC Book Series No. 26, pp. 147-157.

Misra, A.K. and S.C. Mathur. 1983. Sheath rot incidence in cultivars grown at different nitrogen levels. *Int. Rice Res. Newsl.*, **8** (2): 5-6.

Mohan, R. and C.L. Subramanian. 1977a. The effect of certain fungicides on the sheath rot. *Int. Rice Res. Newsl.*, **2**( ): 14.

Mohan, R. and C.L. Subramanian. 1977b. Influence of nitrogen nutrition and sheath rot disease on sugar content of two paddy varieties. *Curr. Sci.*, **45**: 753-755.

Mohan, R. and C.L. Subramanian. 1977c. Studies on the epidemiology of sheath rot disease of rice. *Food Farmg. Agric.*, **9**: 71-72.

Mohan, R. and C.L. Subramanian. 1977d. Some aspects of sheath rot disease of rice caused by *Acrocylindrium oryzae* Sawada, an incitant of sheath rot disease of rice. *Madras Agric. J.*, **66**: 195.

Mohan, R. and C.L. Subramanian. 1978. Growth studies of of *Acrocylindrium oryzae* Sawada an incitant of sheath rot disease of rice. *Madras Agric. J.*, **65**: 172-175.

Mohan, R. and C.L. Subramanian. 1979a. Yield loss due to sheath rot disease of rice caused by *Acrocylindrium oryzae* Sawada. *Madras Agric. J.*, **66**: 195.

Mohan, R. and C.L. Subramanian. 1979b. Influence of nitrogen fertilization on the incidence of sheath rot disease of paddy. *Proc. Ind. Acad. Sci.*, Ser. B, **88**: 249-252.

Mohan, R. and C.L. Subramanian. 1981. Changes in free amino acids due to sheath rot disease in rice. *MAGC Res. Bull.*, **5**: 3-4.

Mondal, A.H. and S.A Miah, 1986. BR3 reaction to multiple disease infection *Int. Rice Res. Newsl.*, **11** (1): 21-22.

Mukherjee, P. and B.P. Singh. 1980. Testing of indigenous rice germplasm against sheath rot by artificial inoculation. *Indian Phytopath.*; **33**: 149.

Mukherjee, P. and A.S. Yadav, 1989. Histopathology and mode of infection of sheath rot (*Sarocladium attenuatum* Gams and Hawksw.) affected rice (*Oryza sativa* L.). *Int. J. Pl. Pathol.* **7** (2) :85-90.

Mukherjee, P., B.P. Singh and F. Rahman. 1980. Testing of indigenous rice germplasm against sheath rot by artificial inoculation. *Indian Phytopath.*, **34**: 287-290.

Muralidhar, B. 1982. Studies on certain physiological and biochemical changes in rice plants as influenced by potassium fertilization and *Acrocylindrium oryzae* Saw. inoculation. M.Sc. (Ag) thesis, Annamalai Univ. India p. 86.

Muralidharan, K. and G. Venkatarao. 1980. Outbreak of sheath rot on rice. *Int. Rice Res. Newsl.*, **5** (5): 7-8.

Naidu, V.D. and P.S. Reddy. 1989. Relationship between tungro (RTV) and sheath rot (ShR) in three rice cultivars. *Int. Rice Res. Newsl.*, **14**(4): 15.

Naik, R. 1976. Incidence of sheath rot in rice-a potential problem for Sambalpur, Orissa. *Int. Rice Res. Newsl.*, **1**(1): 19.

Naik, R. and J.K. Roy. 1975. Occurrence of sheath rot of rice in Sambalpur. *Rice Path. Newsl.*, **2**: 6.

Nair, M.C. and P.K.S. Rajan. 1978. Sheath rot of rice. *Agric. Res. J. Kerala*, **13**: 105-106.

Narain, A 1992. Recent advances of a few minor diseases of rice posing threats *Ind. Mycol. Pl. Path.* **22**(1) : 1-6.

Narasimhan, V., N. Ramadoss, V.V. Sridhar and Abdul. 1994. Using gypsum to manage sheath rot in rice. *Int. Rice Res. Newsl.*, **19** (2): 27-28.

Narasimhan, V., V.V. Sridhar and K.A. Addul. 1993. Efficacy of botanicals in managing sheath rot (ShR) of rice. *Int. Rice Res. Newsl.*, **18** (3): 33.

Narasimhan, V. P. Vidhyasekaran,H.D. Lewin and A. Chandrasekaran. 1990. Effect of foliar application of various salts on rice sheath rot (ShR) disease. *Int. Rice Res. Newsl.*15(1): 29

Omura, S. 1976. The antibiotic cerulenin, a noval tool for biochemistry as an inhibitor of fatty acid synthesis. *Bacteriol Rev.*, **40**: 681-697.

Omura, S. 1981. Cerulenin. In: *Methods in Enzymology*. Lowenstein, J.M. (ed.) **72**: 520-532. Academic Press, Inc. NewYork.

Ou, S.H. 1973. Report to the Goverment of Thailand on blast and other diseases of rice. *F.A.O. (Technical Assistance Programme) Rep.* **28**: 1963.

Ou, S.H. 1985. *Rice disease*, 2nd ed. *Comm. Wealth Agric. Bureaux* Central Sales, Farnham Royal, Slough, UK, p.380.

Padwick, G.W. 1950. *Manual of Rice Diseases. Comm. Wealth Mycol. Inst.*, London, p. 198.

Parvathi. V. 1979. Studies on sheath rot disease of rice. M.Phil., *Dissertation*, Univ. Madras, Madras. p.48.

Pai, A. 1975. Response of some varieties towards sheath rot *Acrocylindrium oryzae* Saw. infection. *Rice Path. Newsl.*, Feb., 18- 19.

Prabhakaran, J., K. Raghunathan and N.N. Prasad. 1974. Occurrence of sheath rot disease of rice caused by *Acrocylindrium oryzae* Sawada, *Annamalai Univ. Agric. Res. Ann.*, **5**: 182-183.

Purkayastha, R.P. and A. Ghosal. 1982. Sheath rot of rice. *Oryza*, **19**: 78-83.

Purkayastha, R.P. and A. Ghosal. 1983. Production of momilactone associated with resistance of rice cultivars to sheath rot disease. *Curr. Sci.*, **52**: 131-132.

Purkayastha, R.P. and A. Ghosal. 1985. Analysis of cross-reactive antigens of *Acrocylindrium oryzae* and rice in relation to sheath rot disease. *Physiol. Pl. Path.*, **27**:245-252.

Purkayastha, R.P., A. Ghosal and S. Biswas. 1983. Momilactone synthesis associated with differential resistance of tall and semi- dwarf rice cvs to sheath rot disease. *Indian Phytopath.*, **36**: 322- 327.

Raghunathan, V. and G. Vijyaraghavan. 1976. Effect of certain fungicides on the *in vitro* growth and enzymatic activities of *Acrocylindrium oryzae* Sawada. *Madras Agric. J.*, **63**: 396-399.

Rahman, M.M., A.K.M. Shajahan and S.A. Miah. 1982. *Echinochloa colona*, an alternative host of *Sarocladium oryzae* causing sheath rot of rice. *Int. Rice Res. Newsl.*, **7** (5): 16-17.

Raina, G.L. and G. Singh. 1980. Sheath rot outbreak in the Punjab. *Int. Rice Res. Newsl.*, **5** (1): 16.

Raina, G.L., G. Singh and S. Sidhu. 1980. Sources of resistance to major rice diseases in the Punjab.*Int. Rice Res. Newsl.*,**5** (3): 6-7.

Rajan, K.M. and P.V. Nair. 1978. Fungicidal control of sheath rot and glume discolouration in rice. *Indian J. Mycol. Pl. Path.*, **8**:85.

Rajan, K.M. and P.V. Nair. 1979. Reaction of certain rice varieties to sheath rot diseases. *Agric. Res. J. Kerala*, **17**: 259-260.

Rajan, K.M. 1980. Control of sheath rot and glume discolouration of rice. *Indian J. Mycol. Pl. Path.*, **10**: 75-76.

Rajan, K.M. 1981. Mechanical separation of paddy seeds as applied to plant pathology. *Int. Rice Res. Newsl.*, **6**(2): 10-11.

Raju, C.A. 1980. Sheath rot of rice. Survival of the pathogen *Sarocladium oryzae* in infected seed and sheaths. *Indian J. Mycol. Pl. Path.*, **10**: 6.

Raju, C.A. and R.A. Singh. 1981a. Development of inoculation technique and varietal reaction to sheath rot of rice. *Philippine Phytopath.*, **17**: 67-73.

Raju, C.A. and R.A. Singh. 1981b. Studies on sheath rot of rice-II. Chemical control. *Pesticides*, **15**: 26-28.

Ramabadran, R. and R. Velazhahan. 1989. *Basic research for crop disease management*. Management of sheath rot of rice (ed. P. Vidhyasekaran). Daya Publishing House, Delhi. pp. 367-375.

Ramakrishnan, T.S. 1971. *Diseases of rice*. ICAR, New Delhi, p.94.

Raychoudhury, S.P. and R.P. Purkayastha. 1980a. Experimental observations on rice sheath rot disease. *Int. Rice Res. Newsl.*, **5** (4): 13-14.

Raychoudhury, S.P. and R.P. Purkayastha. 1980b. Plant hormone affecting susceptibility of dwarf cultivars of rice to sheath rot disease. *Indian Phytopath.*, **33**: 77-82.

Reddy, E.J.S., M.P. Chary and S.M. Reddy. 1983. *In vitro* evaluation of different fungicides in control of sheath rot of rice caused by *Acrocylindrium oryzae*. *Indian J. Mycol. Pl. Path.*, **28**: 552-555.

Reddy, T.C.V., S. Reddy, A.A. Rao and D.K. Rao. 1974. A note on sheath rot of rice from India. *J. Res. Andhra Pradesh Agric. Univ.*, **2 &3**: 136-138.

Reddy, C.S. and A. Ghosh. 1985. Sheath rot incidence and yield losses in rice due to the joint infection of RTV and sheath rot fungus. *Indian Phytopathol.*, **38**: 165-167.

Reddy, C.S. and A. Ghosh. 1993. Rice resistance to sheath rot identified. *Rice Biotech.* Q. **16**

Saccardo, P.A. 1986. Syll. Fung., **4**: 167

Sawada, K. 1922. Descriptive Catalogue of Formosan fungi II. *Rep. Govt. Res. Inst. Dep. Agric. Formosa*, **2**: 1-135.

Sakthivel, N. 1987. Biological control of *Sarocladium oryzae*, (Sawada) Gams & Hawksworth, sheath rot pathogen of rice with seed bacterization with *Psudomonas fluorescens* Mignla. Ph.D. thesis, Univ. of Madras, Madras, p. 100.

Sakthivel, N. and S.S. Gnanamanickam. 1986a.Bacterization of rice with *Psedomonas fluorescens* reduces sheath rot (ShR) infection. *Int. Rice Res. Newsl.* **11**(3): 17-18.

Sakthivel, N. and S.S. Gnanamanickam. 1986b. Isolation and assay for cerulenin produced by the rice sheath rot pathogen *Sarocladium oryzae* (Saw) Gams. *Curr. Sci.* **55**: 17-18.

Saroja, R., M. Suriachandraselvan, J. Venkatakrishnan, K. Nilakantapillai and T.B. Ranganathan. 1987. Field evaluation of rice culture resistance to leaf folder (LF) and sheath rot (ShR). *Int. Rice Res. Newsl.*, **12**(2): 18.

Satyanarayana, K. and C.S. Reddy. 1979. Suggested revision of scoring system for rice sheath rot disease. *Int. Rice Res. Newsl.*, **4** (2): 6.

Sekhar, M.J. and N.N. Prasad. 1988. Quantitative changes in amino acids as influenced by graded doses of potassium and *Sarocladium oryzae* inoculation in rice. *Madras Agric. J.*, **75**: 59- 61.

Shahjahan, A.K.M., Z. Harahap and M.C. Rush. 1977. Sheath rot of rice caused by *Acrocyclindrium oryzae* in Louisiana. *Pl. Dis. Reptr.*, **61**: 307-310.

Shahjahan, A.K.M., M.A. Choudhury and S.I. Akanda. 1986. Growth and virulence of two isolates of *Sarocladium oryzae* causing sheath rot (ShR) in rice. *Int. Rice Res. Newsl.*, **11**(6): 6-7.

Sharma R.C., G.S. Sidhu, T.S. Bharaj and H.L. Sharma. 1993. Pathological constraints on hybrid rice production technology. *Int. Rice Res. Newsl.*, **18** (3): 15.

Singh, A.P., S.M. Asghar and S.M. Ghufran. 1985. Assessment of yield loss due to sheath rot disease of rice. *Indian Phytopath.* **38** (4) :753-755

Singh, R.A. and C.A. Raju. 1981. Studies on sheath rot of rice. *Int. Rice Res. Newsl.*, **6** (2): 11-12.

Singh, R.N. 1988. Status of sheath rot(ShR) in eastern Uttar Pradesh (U.P.). India, *Int. Rice Res. Newsl.*, **13** (6): 33.

Singh, N. 1982. Carbon and nitrogen nutrition of *Acrocylindrium oryzae* and production of pectic and celluolytic enzymes. *Trans. Br. Mycol. Soc.*, **78**: 355-356.

Singh, B.N., S.P. Sahu, Y. Prasad and R.S. Singh. 1986. Seedling age in relation to sheath rot (ShR) occurrence in rice. *Int. Rice Res. Newsl.*, **11**(4): 26.

Sreekumar, S.G., V. Gopinathnair, A.R. Balakrishnan and P. Manikantannair. 1992. Genetic evaluation of pre-release cultures varieties of rice for yield to sheath blight and sheath rot diseases. *Agric. Res. J. Kerala* **30** (1): 46-49.

Srinivasan, S. 1980a. Yield losses caused by sheath rot. *Int. Rice Res. Newsl.*, **5** (2): 4.

Srinivasan, S. 1980b. Effect of nitrogen and spacing on sheath rot incidence in rice. *Int. Rice Res. Newsl.*, **5** (5): 15-16.

Srinivasan, S. 1981. Seasonal incidence of sheath rot in Thanjavur Delta, India. *Int. Rice Res. Newsl.*, **6** (5): 16.

Srinivasan, S. and A. Pari. 1978. Sheath rot (*Acrocylindrium oryzae*). *Aaduthurai Rept.*, **2**: 68.

Subramanian, C.L. and G. Ramakrishnan. 1975. Varietal reactions to the sheath rot of rice (*Acrocylindrium oryzae* Saw.) *Curr. Sci.*, **44**: 405- 406.

Surin, A. and S. Disthaporn. 1977. 'Rice abortion' or sheath rot- a serious rice disease in Thailand. *Int. Rice Res. Newsl.*, **2** (5): 16- 17.

Surin, A., P. Arunyanart, R. Dhitikiattipong, W. Rojanahusdin, S. Disthaporn and K. Soontrajan. 1988. Rice yield loss to sheath rot (ShR). *Int. Rice Res. Newsl.*, **13**(6): 34.

Tasugi, H. and Y. Ikeda. 1956. Studies on the sheath rot of rice plants caused by *Acrocylindrium oryzae* Sawada. *Bull. Nat. Inst. Agric. Sci. Tokyo, Ser. C.*, **6**: 151-166.

Tikoo, M.L. 1985. Effect of sheath rot (*Acrocylindrium oryzae*) on yield components of rice. *Rice Dev. Rept.*, **2**: 24-25.

Thrimurty, V.S., O.P. Veda, R.G. Satputhe and P. Kashyap. 1980. Sheath rot incidence and chaffy grain percentage on some popular rice. *Int. Rice Res. Newsl.*, **5** (5): 7.

Tombisana Devi R.K., N. Iboton and N. Singh. 1994. Effect of temperature and light on growth and sporulation of fusarium rice sheath rot. *Int. Rice Res. Newsl.*, **19** (3): 28.

Tripathi, R.S. 1982. Occurrence of sheath rot in Rajasthan. *Int. Rice Res. Newsl.*, **7** (1): 5.

Turner. W.B. and D.C. Aldridge. 1983. *Fungal Metabolites* II. Academic Press London p. 109.

Upadhya, R.K. Rice disease status in India. 1985. *Int. Rice Res. Newsl.*, **10** (5): 17-18.

Upadhyay, R.K. and M.C. Diwakar. 1984. Sheath rot (ShR) in Chatisgarh, Madhya Pradesh, India. *Int. Rice Res. Newsl.*, **9** (5): 6.

Vasantha Devi, T. and S.S. Gnanamanickam. 1993. Toxin produced by *Sarocladium oryzae* involved in inducing sheath rot (ShR) symptoms in rice. *Int. Rice Res. Newsl.*, **18** (4): 30-31.

Vasavan, M.G., K.M. Rajan and M.J. Thomas. 1980. Herbicides in plant disease control. *Int. Rice Res. Newsl.*, **5** (4): 18.

Venkataraman, S. A. Ghosh and R.K. Mahajan. 1987. Synergistic effect of rice tungro virus and *Sarocladium oryzae* on sheath rot disease of rice. *Int. Pl. Dis.*, **5**: 141-145.

Vidhyasekaran, P., K. Ranganathan, B. Rajamanickam and J. Radhakrishan. 1984. Quality of rice grains from sheath rot-affected plants. *Int. Rice Res. Newsl.*, **9**(1): 19.

Viswanathan, R. and P. Narayanasamy. 1990. Potential antagonist of rice sheath rot ShR pathogen. *Int. Rice Res. Newsl.* **15** (2): 26-27.

Wei, C. T. 1957. *Manual of rice pathogens*. Science press, Beijing. p. 267.

Yosida, S., D.A. Farno, J.H. Cock and K.A. Gomez. 1976. *Laboratory manual for physiological studies of rice* (3rd ed.). *Int. Rice Res. Inst.* (IRRI), Box. 933. Manila, Philippines.

Zeigler, R.S. and E. Alrarez. 1989. *Pseudomonas* spp. causing rice sheath rot (ShR) and grain discoloration (GID). *Int. Rice Res. Newsl.*, **14**(1): 26.

Index

[Note : Rice cultivars and ShR pathogens are not listed]